AF831422

Busy Ant Maths

Test Pack 3

Series Editor: Peter Clarke

Author: Caroline Fawcus

William Collins' dream of knowledge for all began with the publication of his first book in 1819. A self-educated mill worker, he not only enriched millions of lives, but also founded a flourishing publishing house. Today, staying true to this spirit, Collins books are packed with inspiration, innovation and practical expertise. They place you at the centre of a world of possibility and give you exactly what you need to explore it.

Collins. Freedom to teach.

Published by Collins
An imprint of HarperCollins*Publishers*
The News Building
1 London Bridge Street
London
SE1 9GF

HarperCollins Publishers
Macken House
39/40 Mayor Street Upper
Dublin 1
D01 C9W8
Ireland

Browse the complete Collins catalogue at
www.collins.co.uk

10 9 8 7 6 5 4 3 2

ISBN 978-0-00-816738-7

British Library Cataloguing in Publication Data
A Catalogue record for this publication is available from the British Library.

Author: Caroline Fawcus
Series Editor: Peter Clarke
Commissioning Editor: Fiona McGlade
Project Editor: Leah Willey
Cover design and artwork: Amparo Barrera
Internal design and illustrations: Steve Evans and Mike Connor
Production: Robin Forrester

MIX
Paper | Supporting responsible forestry
FSC™ C007454

This book contains FSC™ certified paper and other controlled sources to ensure responsible forest management.

For more information visit: www.harpercollins.co.uk/green

Contents

Introduction

Test Papers – Pupil version

Test Papers – Teacher version

Introduction

Purpose of the Busy Ant Maths Year 3 Test Pack

1. To assist teachers in assessing the overall level of mastery achieved by pupils in the mathematics programme of study for Year 3, and in determining pupils' readiness to deal with the expectations of the Year 4 programme of study.
2. To provide parents, teachers and school leaders with information on how pupils are performing in comparison with national standards, by assigning one of the following:
   - working below national standard
   - working towards national standard
   - working at national standard
   - working at greater depth within the national standard.
3. To enable teachers and school leaders to monitor the performance of pupil cohorts, to identify where interventions may be required and to ensure that pupils are supported to achieve sufficient progress and expected attainment.

The Busy Ant Maths Test Packs are intended to be used as summative assessments towards the end of the academic year. They are not designed to assess pupils' level of mastery in each of the individual National Curriculum attainment targets and domains. For guidance on making formative assessments regarding pupils' strengths and weaknesses and level of mastery in specific attainment targets and domains, teachers should refer to the Assessment Tasks and Assessment Exercises in the Busy Ant Maths Assessment Guides.

Structure of the Busy Ant Maths Year 3 Test Pack

Paper	Paper 1: arithmetic	Paper 2: reasoning	Paper 3: reasoning
Total marks	30 marks	30 marks	30 marks
Recommended timing	25 minutes	30 minutes	30 minutes
National Curriculum domain coverage	Number – Number and place value Number – Addition and subtraction Number – Multiplication and division Number – Fractions		
		Measurement Geometry – Properties of shapes Statistics	

The papers should be administered in order.

Pupils can have a break between the papers.

Administering Paper 1: arithmetic

Recommended timing

Pupils have 25 minutes to complete the paper.

Resources

Pupils will need the following resources:

- a blue or black pen, or a dark pencil
- a rubber (optional). If rubbers are not provided, tell pupils that they may cross out any answer they wish to change.

Pupils may also use the following resources, if this is normal classroom practice, provided they only give word-for-word translations:

- bilingual dictionaries or electronic translators
- bilingual word lists
- monolingual English electronic spell checkers.

Pupils are not allowed:

- calculators
- other mathematical equipment, such as angle measurers, mirrors or rulers.

Assistance

Teachers should ensure that nothing they do or say during the test could be interpreted as giving pupils an advantage.

If a pupil requests it, teachers may read a question out loud on a one-to-one basis.

If reading to a pupil, teachers should read words and numbers but not mathematical symbols. This is to ensure that pupils are not given an unfair advantage by having the operation/function inadvertently explained by reading its name.

Before the test begins

- Review the list of pupils with any particular individual needs and ensure that their requirements are met.
- Ensure that pupils are able to work undisturbed and that the classroom layout and seating arrangements are suitable.
- Ensure that any wall displays do not give pupils an unfair advantage.
- Write the start and finish times of the test on the board.
- Provide each pupil with the resources they require.

What to say at the start of the test

- *This is the arithmetic paper.*
- *You will need a blue or black pen or a dark pencil. You may use an rubber for this test.* (If rubbers are not provided, tell the pupils that they may cross out any answer that they wish to change.)
- *Write your name and class name/number on the front of your test paper.*
- *Look at the list of instructions on page 1. I will read the instructions to you.* (Read the instructions on page 1 of the test paper to the pupils.)

Instructions

You **may not** use a calculator to answer any questions in this test.

Questions and answers

You have **25 minutes** to complete this test.

Work as quickly and as carefully as you can.

Put your answer in the box for each question.

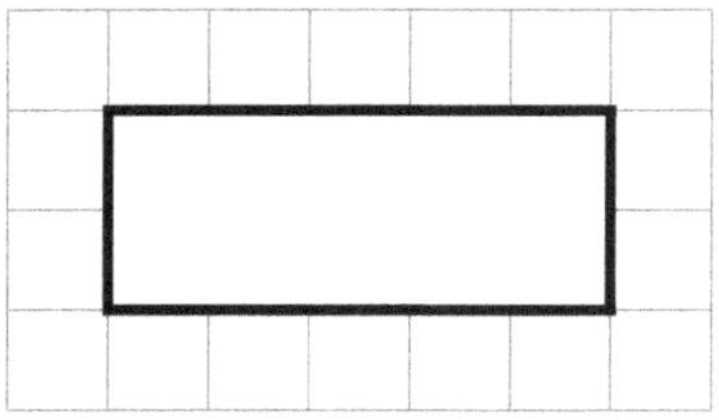

For questions expressed as common fractions, you should give your answer as common fractions.

All other answers should be given as whole numbers.

If you cannot do one of the questions, **go on to the next one.** You can come back to it later if you have time.

If you finish before the end, **go back and check your work**.

Marks

The number under each box at the side of the page tells you the maximum number of marks for each question.

In this test, some addition, subtraction and multiplication questions are worth **2 marks each.** You will be awarded 2 marks for a correct answer. You may get 1 mark for showing a formal method.

All other marks are worth **1 mark each**.

- *If you want to change your answer, put a line through the response you don't want, or use a rubber.*
- *Remember to check your work carefully.*
- *If you have any questions during the test, put your hand up and wait for someone to come to you. Remember, I can't help you answer any of the test questions.*
- *You must work on your own and you may not talk to each other.*
- *Are there any questions you want to ask me now?*
- *I will tell you when you have 5 minutes left.*
- *I will tell you when the test is over and to stop writing.*
- *You may now start the test.*

Administering Papers 2 and 3: reasoning

Recommended timing

Pupils have 30 minutes to complete each paper.

Resources

Pupils will need the following resources for each test:

- a blue or black pen, or a dark pencil
- a rubber (optional). If rubbers are not provided, tell the pupils that they may cross out any answer they wish to change.
- a sharp, dark pencil for mathematical drawing
- a ruler (showing centimetres and millimetres).

Pupils may also use the following resources, if this is normal classroom practice, provided they only give word-for-word translations:

- bilingual dictionaries or electronic translators
- bilingual word lists
- monolingual English electronic spell checkers.

Pupils are not allowed:

- calculators.

Assistance

Teachers should ensure that nothing they do or say during the tests could be interpreted as giving pupils an advantage.

If a pupil requests it, teachers may read a question out loud on a one-to-one basis.

If reading to a pupil, teachers should read words and numbers but not mathematical symbols. This is to ensure that pupils are not given an unfair advantage by having the operation/function inadvertently explained by reading its name.

At a pupil's request, teachers can point to parts of the test paper such as charts, diagrams, statements and equations, but they should not explain the information or help the pupil by interpreting it.

If any everyday context or words related to a question are unfamiliar to a pupil, teachers may show them related objects or pictures, or describe the related context.

Before each test begins

- Review the list of pupils with any particular individual needs and ensure that their requirements are met.
- Ensure that pupils are able to work undisturbed and that the classroom layout and seating arrangements are suitable.
- Ensure that any wall displays do not give pupils an unfair advantage.
- Write the start and finish times of the test on the board.
- Provide each pupil with the resources they require.

What to say at the start of the test

- *This is the first/second reasoning paper.*
- *You will need a blue or black pen or dark pencil, a sharp pencil for mathematical drawing, and a ruler. You may use a rubber for this test.* (If rubbers are not provided, you should tell the pupils that they may cross out any answer that they wish to change.)
- *Write your name and class name/number on the front of your test paper.*

- *Look at the list of instructions on page 1. I will read the instructions to you.* (Read the instructions on page 1 of the test paper to the pupils.)

Instructions

You **may not** use a calculator to answer any questions in this test.

Questions and answers

You have **30 minutes** to complete this test.

Follow the instructions for each question.

Work as quickly and as carefully as you can.

If you need to do working out, you can use the space around the question.

Some questions have a method box like this:

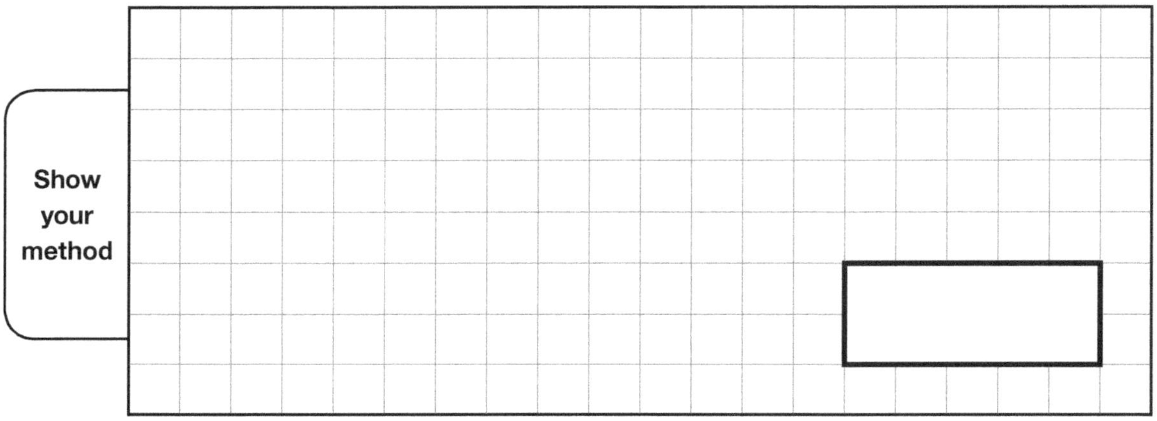

For these questions you may get a mark for showing your method.

If you cannot do one of the questions, **go to the next one.**
You can come back to it later, if you have time.

If you finish before the end, **go back and check your work**.

Marks

The number under each line at the side of the page tells you the maximum number of marks for each question.

- *If you want to change your answer, put a line through the response you don't want, or use a rubber.*
- *Remember to check your work carefully.*
- *If you have any questions during the test, put your hand up and wait for someone to come to you. Remember, I can't help you answer any of the test questions.*
- *You must work on your own and you may not talk to each other.*

- *Are there any questions you want to ask me now?*
- *I will tell you when you have 5 minutes left.*
- *I will tell you when the test is over and to stop writing.*
- *You may now start the test.*

What to do at the end of the test

- Use the mark schemes to mark the three papers, following the '**General marking guidance**', the '**Requirements**', and any '**Additional guidance**' for each question.
- Write the total score for each paper on the front of the test paper.
- Calculate the total marks allocated for all three papers to determine the pupil's overall score out of 90. Record their results.
- If you wish to compare pupils' achievements with national standards, refer to the **Assigning a National standard** table at the end of this Introduction.

Explanation of the mark schemes

The marking information for each question is set out in the form of tables which start on page xiii.

The '**Qu.**' column on the left-hand side of each table provides a quick reference to the question number and, where applicable, the question part.

The '**Requirement**' column may include two types of information:

- A statement of the requirements for the award of each mark, with an indication of whether credit can be given for a correct method.
- Examples of some different types of correct response.

The '**Mark**' column indicates the total number of marks available for each question part.

The '**Additional guidance**' column indicates alternative acceptable responses, and provides details of specific types of response that are unacceptable. Other guidance, such as the range of acceptable answers, is provided as necessary.

General marking guidance

What if...	Marking procedure
The pupil's response is numerically equivalent to the answer in the mark scheme.	Award the mark unless the mark scheme states otherwise.
The pupil's response does not match closely any of the examples given.	Teachers should use their professional judgement in deciding whether the response corresponds with the statement of the requirements given in the 'Requirement' column. Consideration should also be given to the 'Additional guidance' column.
The pupil has responded in a non-standard way.	Pupils may provide evidence in a form as long as its meaning can be understood. Diagrams, symbols or words are acceptable for explanations or for indicating a response. In Paper 1: arithmetic, pupils should use formal methods for calculating their answers. For those questions that are worth 2 marks, a partial credit of 1 mark should be awarded for evidence of using formal methods with one arithmetic error. In Papers 2 and 3: reasoning, a partial mark (or marks) should be awarded for evidence of a complete and correct method.
There appears to be a misreading affecting the working.	If the original intention or difficulty level of the question is not reduced deduct 1 mark only. In 1-mark questions, 0 marks are awarded. In 2-mark questions that have a method mark, 1 mark should be awarded if the correct method is correctly implemented with the misread number. If the original intention or difficulty level of the question is reduced do not award any part for the question part.
No answer is given in the expected place, but the correct answer is given elsewhere.	Where a pupil has shown understanding of the question, the mark(s) should be given.
The pupil's answer is correct but the wrong working is shown.	A correct response should always be marked as correct unless the mark scheme states otherwise.
The pupil has worked out the answer correctly and then written an incorrect answer in the answer box.	Precedence should be given to the answer given in the answer box over any other workings. There may be cases where the incorrect answer is due to a transcription error. In such cases check the pupil's intention and decide whether to award the mark.
The correct response has been crossed (or rubbed) out and not replaced.	Any legible crossed-out work that has not been replaced should be marked according to the mark schemes. If the work is replaced, then crossed-out work should not be considered.
More than one answer is given.	If all answers are correct (or a range of answers is given, all of which are correct), the mark should be awarded unless prohibited by the mark schemes. If both correct and incorrect responses are given, no mark should be awarded unless the mark scheme states otherwise.
The answer is correct but, in a later part of the question, the pupil has contradicted this response.	A mark given for one part should not be disallowed for working or answers given in a different part, unless the mark scheme specifically states otherwise.

Responses involving money

	Accept	Do not accept
Where the £ is given: £	£1.60 £5 £5.00 Any unambiguous indication of the correct amount, e.g. £1.60p £1 60 pence £1 60 £1,60 £1-60 £1:60	Incorrect placement of pounds or pence, e.g. £160 £160p Incorrect placement of decimal point, or incorrect use or omission of 0, e.g. £1.6 £1 600 £16 0 £1-6-0
Where the p is given: p	20p Any unambiguous indication of the correct amount, e.g. £0.20p	Incorrect or ambiguous use of pounds or pence, e.g. 0.20p £20p
Where no sign is given:	£1.60 160p 20p £0.20 Any unambiguous indication of the correct amount, e.g. £1.60p £1 60 pence £1 60 £1,60 £1-60 £1:60 1.60 160 1 pound 60 £0.20p £.20p £.20 20 0.20	Incorrect or ambiguous use of pounds or pence, e.g. £160 £160p £1.6 1.60p £20 £20p 0.2 0.20p

Responses involving time

	Accept	Do not accept
A time interval, e.g. 1 hour 30 minutes	1 hour 30 minutes Any unambiguous, correct indication, e.g. $1\frac{1}{2}$ hours 1·5 hours 1h 30 1h 30 min 1 30 90 minutes 90 Digital electronic time, i.e. 1:30	Incorrect or ambiguous time interval, e.g. 1·30 1-30 1,30 130 1·3 1·3 hours 1·3h 1h 3 1·30 min
A specific time, e.g. 3:20 a.m., 15:20	3:20 a.m. 3:20 twenty past three Any unambiguous, correct indication, e.g. 03·20 3·20 0320 3 20 3-20 3,20 Unambiguous change to 12- or 24-hour clock, e.g.: 18:10 as 6:10 p.m. or 18:10 p.m.	Incorrect time, e.g. 3·2 a.m. 3·20 p.m. Incorrect placement of separators, spaces, etc. or incorrect use or omission of 0, e.g. 320 3:2:0 3·2 032

Responses involving measures

	Accept	Do not accept
Where units are given, e.g. m, kg, l m	1·5 m Any unambiguous indication of the correct measurement, e.g. $1\frac{1}{2}$ m 1·50 m 1 m 50 cm	Incorrect or ambiguous use of units, e.g. 1500 m

Notes:

- If a pupil leaves the answer box empty but writes the answer elsewhere on the page, then that answer must be consistent with the units given in the answer box and the conditions listed above.
- If a pupil changes the unit given in the answer box, then their answer must be equivalent to the correct answer using the unit they have chosen, unless otherwise indicated in the mark schemes.

Mark schemes for Paper 1: arithmetic

Qu.	Requirement	Mark	Additional guidance
1	77	1	
2	37	1	
3	9	1	
4	618	1	
5	355	1	
6	476	1	
7	67	1	
8	187	1	
9	221	1	
10	36	1	
11	164	1	
12	$\frac{3}{5}$	1	Accept equivalent fractions, or an exact decimal equivalent, e.g. 0·6. As the question is expressed in common fractions, pupils should give their answer as a common fraction. An equivalent fraction such as $\frac{6}{10}$ should also be awarded the mark. Since this fraction does have an exact decimal equivalent, the mark scheme also allows this to be awarded the mark.
13	4	1	
14	9	1	
15	27	1	
16	600	1	
17	111	1	
18	11	1	
19	96	1	
20	916	1	
21	20	1	

Qu.	Requirement	Mark	Additional guidance
22	$\frac{5}{9}$	**1**	Accept exact decimal equivalent, i.e. $0{\cdot}\dot{5}$ or $0{\cdot}\bar{5}$. (**Do not** accept rounded or truncated decimals.) Accept equivalent fractions, e.g. $\frac{10}{18}$. As the question is expressed in common fractions, pupils should give their answer as a common fraction. An equivalent fraction such as $\frac{10}{18}$ should also be awarded the mark. The fraction does have a recurring decimal equivalent which should also be awarded the mark. However, a decimal answer truncated to 0.6 or rounded to 0.56 for example should not be awarded the mark.
23	Award **TWO** marks for the correct answer of 924 If the answer is incorrect, award **ONE** mark for the formal addition method which contains no more than **ONE** arithmetical error, e.g. $\begin{array}{r} 2\ \ 5\ \ 6 \\ +\ \ 6\ \ 6\ \ 8 \\ \hline \text{Wrong answer} \\ \hline {\scriptstyle 1\ \ \ 1\ \ } \end{array}$	**Up to 2**	**TWO** marks are awarded for the correct answer. However, if the answer is incorrect, one mark can only be awarded if the pupil has used the formal method of addition. **Do not** award any marks if there is no evidence of the units being carried into the tens column, and the tens being carried into the hundreds column.
24	Award **TWO** marks for the correct answer of 216 If the answer is incorrect, award **ONE** mark for the formal method of short multiplication which contains no more than **ONE** arithmetical error, e.g. $\begin{array}{r} 2\ \ 4 \\ \times\ \ \ \ \ 9 \\ \hline \text{Wrong answer} \\ \hline {\scriptstyle 3\ \ \ \ \ } \end{array}$	**Up to 2**	**TWO** marks are awarded for the correct answer. However, if the answer is incorrect, one mark can only be awarded if the pupil has used the formal method of short multiplication. **Do not** award any marks if there is no evidence of the units being carried into the tens column.
25	Award **TWO** marks for the correct answer of 576 If the answer is incorrect, award **ONE** mark for a formal subtraction method which contains no more than **ONE** arithmetical error, e.g. $\begin{array}{r} {}^{6}\not{7}\ {}^{14}\not{5}\ {}^{1}3 \\ -\ \ 1\ \ 7\ \ 7 \\ \hline \text{Wrong answer} \\ \hline \end{array}$ Or $\begin{array}{r} 7\ {}^{1}5\ {}^{1}3 \\ -\ \ \underset{2}{\not{1}}\ \ \underset{8}{\not{7}}\ \ 7 \\ \hline \text{Wrong answer} \\ \hline \end{array}$	**Up to 2**	**TWO** marks are awarded for the correct answer. However, if the answer is incorrect, one mark can only be awarded if the pupil has used a formal method of subtraction. **Do not** award any marks if there is no evidence of using a formal method, e.g. for subtraction by decomposition, one ten being exchanged for ten ones, and one hundred being exchanged for ten tens.
26	32	**1**	
27	20	**1**	

Mark schemes for Paper 2: reasoning

Qu.	Requirement	Mark	Additional guidance
1	600	1	
2a 2b	3 Grace	1 1	Do **not** accept a list of names. Accept abbreviations or misspellings if the intention is clear, e.g. G.
3	Line drawn that measures in the range 5·8 cm and 6·2 cm inclusive.	1	Do **not** award the mark if a ruler has not been used. Pupils are expected to use a ruler to draw a line of exactly 6 cm. A tolerance of ± 2 mm is allowed.
4	415	1	Do **not** award the mark for 400 10 5
5	807	1	
6	2	1	Do **not** accept 180
7	509	1	
8	717 – 40 = 687 crossed, as shown. 217 + 8 = 225 179 + 400 = 579 827 – 600 = 227 717 – ✗ = 687 509 – 20 = 489	1	Pupils are expected to add and subtract ones, tens and hundreds to and from a three-digit number mentally. Pupils will be awarded the mark for any method used, written or mental, although pupils that choose a written method are likely to spend longer on the question.
9	974	1	
10	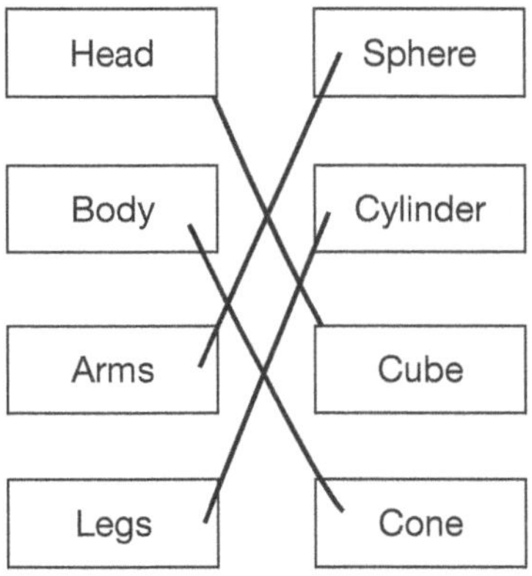	1	**Do not** award the mark if the answer given is not a one-to-one mapping.
11	6 × 8 and 4 × 12 circled as shown. 4 × 8 6 × 8 (circled) 4 × 12 (circled) 3 × 12 8 × 12 6 × 12 3 × 4	1	Both calculations must be circled, with no extras.
12	Answers in the range 20–24 inclusive.	1	Pupils are expected to estimate numbers using different representations.

Qu.	Requirement	Mark	Additional guidance
13	The square and rectangle ticked, as shown. 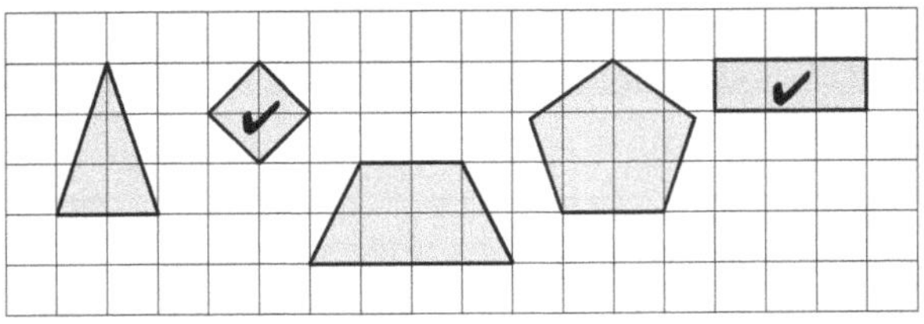	1	Both shapes must be indicated for an award of **ONE** mark. **Do not** award the mark if more than 2 shapes have been indicated.
14	$\frac{6}{10}$ and $\frac{8}{10}$	1	Fractions must be in the correct order. Accept any equivalent fractions, e.g. $\frac{3}{5}$ and $\frac{4}{5}$ **Do not** accept decimal equivalents.
15	20 cm	1	
16	Any 4 bees circled, as shown. 	1	
17	Hands drawn as shown. 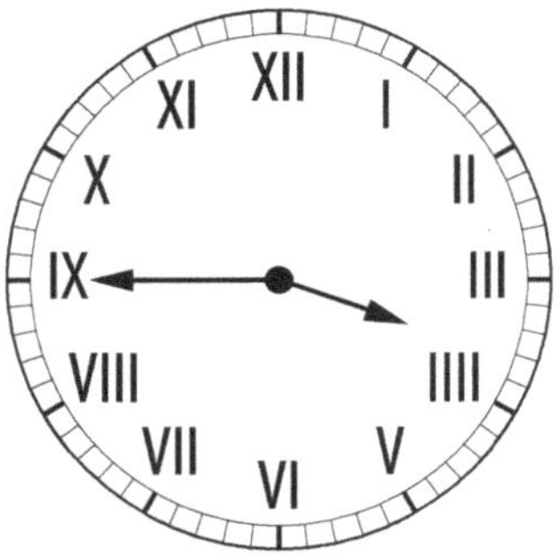	1	
18	7	1	
19	An arrow drawn to 460g, as shown. 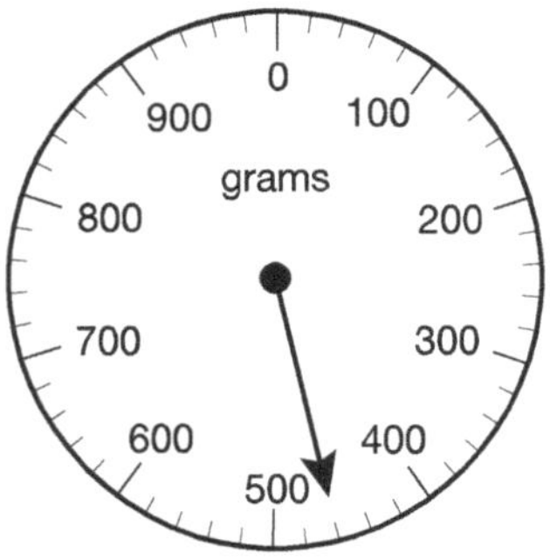	1	
20	$\frac{2}{8}$	1	Accept equivalent fractions and decimals, e.g. $\frac{1}{4}$ or 0·25

Qu.	Requirement	Mark	Additional guidance
21	Wednesday	1	Accept We or W. Pupils are expected to use their knowledge of the number of days in each month to answer this question.
22	5	1	
23	More than 1 litre ticked, as shown. Less than 1 litre ☐ Exactly 1 litre ☐ More than 1 litre ☑	1	**Do not** award the mark if more than one box has been ticked. Pupils are expected to make simple comparisons between measures, including mixed units.
24	4	1	
25	Award **TWO** marks for a correct answer of £96 If the answer is incorrect, award **ONE** mark for evidence of an appropriate method, e.g. $3 \times 4 \times 8$	Up to 2	Answer need not be obtained for the award of **ONE** mark.
26	$\frac{1}{12}$	1	
27	Award **TWO** marks for all five digits correctly placed, as shown. $\boxed{4}\boxed{2} + 6\boxed{3} = \boxed{1}\boxed{0}5$ or $\boxed{4}\boxed{3} + 6\boxed{2} = \boxed{1}\boxed{0}5$ If the answer is incorrect, award **ONE** mark for either the 2 and 3 correctly placed, or award **ONE** mark for the 1, 0 and 4 correctly placed.	Up to 2	

Mark schemes for Paper 3: reasoning

Qu.	Requirement	Mark	Additional guidance
1	600	**1**	
2	Any 6-sided enclosed shape, for example: 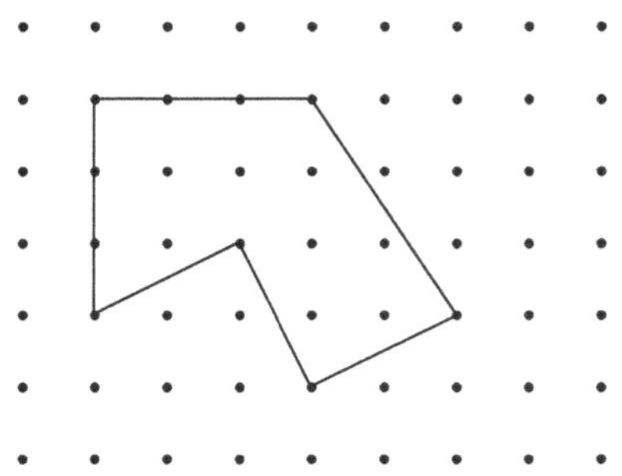	**1**	Accept inaccurate drawing provided the intention is clear.
3a **3b**	243 836	**1** **1**	
4	209, 279, 290, 309, 319	**1**	
5	Answers in the range 81–85 inclusive.	**1**	Answers must be given in millimetres for the award of the mark.
6	Award **TWO** marks for the correct answer of 255 If the answer is incorrect, award **ONE** mark for evidence of an appropriate method, e.g. 800 – 545	**Up to 2**	
7	9:05 p.m.	**1**	Accept 5 past 9 p.m. or any similar correct writing of the time as long as p.m. notation has been used.
8	8 **AND** 8	**1**	
9	225 ml	**1**	
10	3	**1**	
11	Correct shape ticked, as shown. 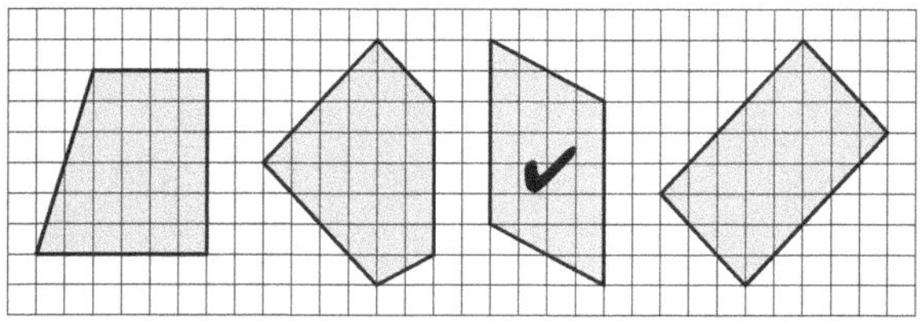	**1**	
12	Tennis racket ticked as shown. 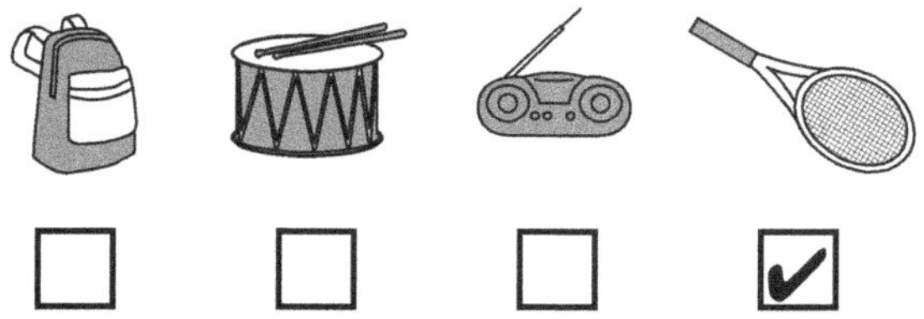	**1**	**Do not** award the mark if more than one box is ticked.

Qu.	Requirement	Mark	Additional guidance
13	10	1	
14	D	1	
15	Arrow drawn as shown.  	1	
16	Ana and Caz ticked as shown. Ana [✔] Barnaby [ ] Caz [✔] Dinesh [ ] Eva [ ]	1	**Do not** award the mark if extra boxes are ticked.
17a 17b	30 11	1 1	
18	Correct shape ticked, as shown. 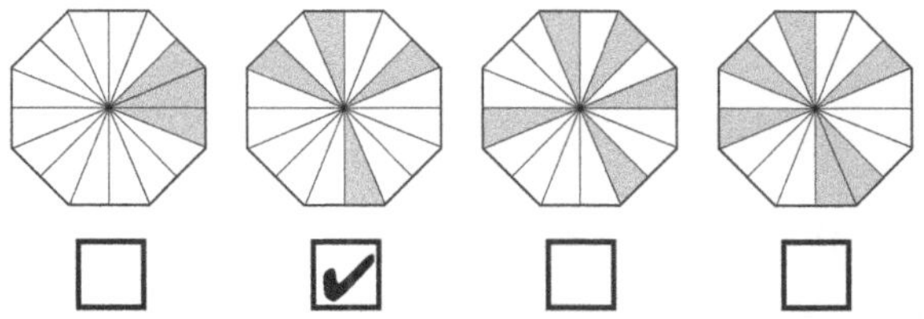	1	**Do not** award the mark if more than one box is ticked.
19	325 385 (345) 440 (355) 360	1	
20	Award **TWO** marks for the correct answer of 24p If the answer is incorrect, award **ONE** mark for evidence of an appropriate method, e.g. 38p × 2 = 76p £1 – 76p Or £1 – 38p = 62p 62p – 38p	Up to 2	Accept any unambiguous indication of the correct amount, e.g. £0.24p. Answer need not be obtained for the award of **ONE** mark.
21	111	1	
22	All boxes completed correctly, as shown. $\frac{1}{3}$ [>] $\frac{1}{6}$ $\frac{4}{3}$ [>] 1 $\frac{2}{9}$ [<] $\frac{5}{9}$	1	
23	504	1	

Qu.	Requirement	Mark	Additional guidance
24	$\frac{2}{5}$	1	**Do not** accept decimal equivalents
25	Award **TWO** marks for the correct answer of 128 If the answer is incorrect, award **ONE** mark for an appropriate method, e.g. 5 + 3 = 8 8 × 16 Or 5 × 16 = 80 3 × 16 = 48 80 + 48	Up to 2	Answer need not be obtained for the award of **ONE** mark.

Assigning a national standard

A total of 90 marks are available for the Year 3 Test:

- Paper 1: arithmetic (30 marks)
- Paper 2: reasoning (30 marks)
- Paper 3: reasoning (30 marks).

The sum of the marks allocated from these two papers determines the pupil's overall raw score.

Schools that wish to compare their pupils' achievements with national standards should use the thresholds in the table below to convert raw scores into national standards.

Raw score (out of 90)	0–17	18–44	45–71	72–90
Scaled score	0%–19%	20%–49%	50%–79%	80%–100%
Standard	Working below national standard	Working towards national standard	Working at national standard	Working at greater depth within the national standard.

Year 3
PAPER
1

Arithmetic

Name: ______________________

Class: ______________________

Instructions

You **may not** use a calculator to answer any questions in this test.

Questions and answers

You have **25 minutes** to complete this test.

Work as quickly and as carefully as you can.

Put your answer in the box for each question.

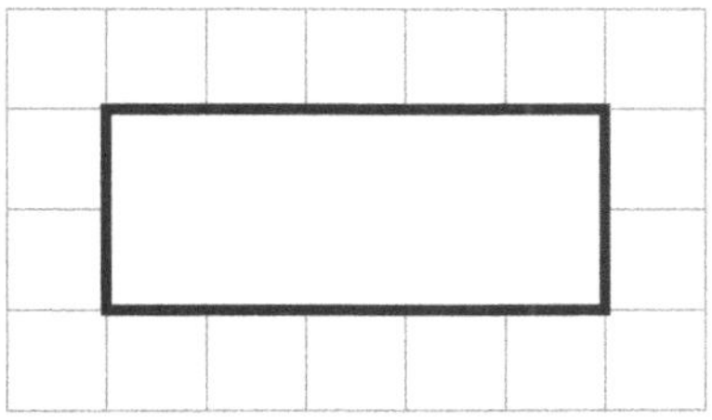

For questions expressed as common fractions, you should give your answer as common fractions.

All other answers should be given as whole numbers.

If you cannot do one of the questions, **go on to the next one**. You can come back to it later if you have time.

If you finish before the end, **go back and check your work**.

Marks

The number under each box at the side of the page tells you the maximum number of marks for each question.

In this test, some addition, subtraction and multiplication questions are worth **2 marks each**. You will be awarded 2 marks for a correct answer. You may get 1 mark for showing a formal method.

All other marks are worth **1 mark each**.

1 $67 + 10 =$

1 mark

2 $44 - 7 =$

1 mark

3 $3 \times 3 =$

1 mark

4 658 − 40 =

1 mark

5 55 + 300 =

1 mark

6 576 − 100 =

1 mark

7 52 + 15 =

1 mark

8 178 + 9 =

1 mark

9 161 + 60 =

1 mark

10 $9 \times 4 =$

1 mark

11 $365 - 201 =$

1 mark

12 $\frac{2}{5} + \frac{1}{5} =$

1 mark

13 $16 \div 4 =$

1 mark

14 $\frac{1}{2}$ of $18 =$

1 mark

15 $76 - 49 =$

1 mark

16 $550 + 50 =$

1 mark

17 $32 + 79 =$

1 mark

18 $33 \div 3 =$

1 mark

19 $12 \times 8 =$

1 mark

20 $854 + 62 =$

1 mark

21 $80 \div 4 =$

1 mark

22

$\frac{7}{9} - \frac{2}{9} =$

1 mark

23

$$\begin{array}{r} 2\,5\,6 \\ +\;6\,6\,8 \\ \hline \end{array}$$

Show your method

2 marks

24

$$\begin{array}{r} 2\,4 \\ \times\;\;9 \\ \hline \end{array}$$

Show your method

2 marks

25

$$\begin{array}{r} 753 \\ -\ 177 \\ \hline \end{array}$$

Show your method

2 marks

26

$96 \div 3 =$

1 mark

27

$\frac{4}{5}$ of 25 =

1 mark

Year 3
PAPER 2

Reasoning

Name: ______________________

Class: ______________________

Instructions

You **may not** use a calculator to answer any questions in this test.

Questions and answers

You have **30 minutes** to complete this test.

Follow the instructions for each question.

Work as quickly and as carefully as you can.

If you need to do working out, you can use the space around the question.

Some questions have a method box like this:

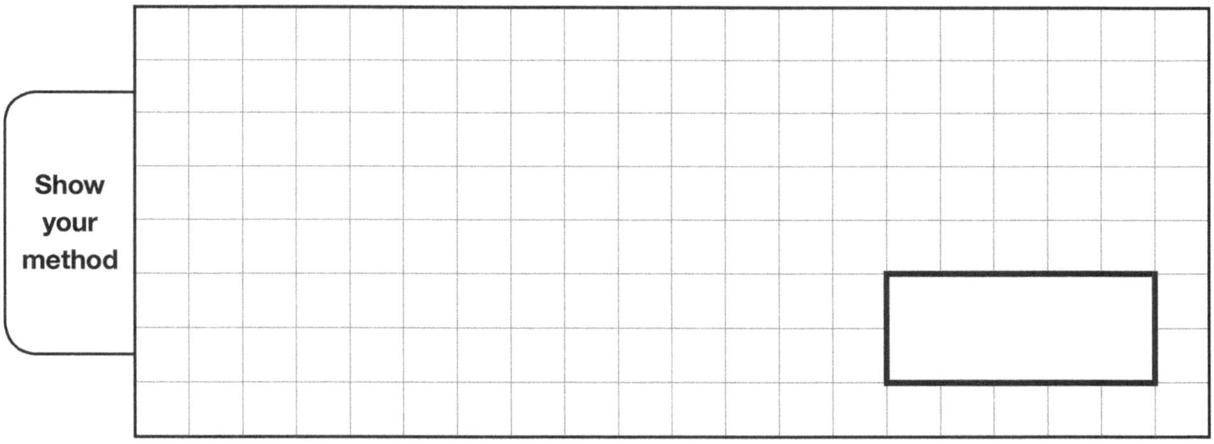

For these questions you may get a mark for showing your method.

If you cannot do one of the questions, **go to the next one**.
You can come back to it later, if you have time.

If you finish before the end, **go back and check your work**.

Marks

The number under each line at the side of the page tells you the maximum number of marks for each question.

1 Write the missing number.

400	500		700

1 mark

2 Children can attend playgroup for the morning, lunch and afternoon.

The sessions attended by 6 children are shown in the table.

	Morning	Lunch	Afternoon
Ava	✔	✔	
Grace	✔	✔	✔
Jayden		✔	✔
Kai	✔		
Frankie			✔
Yasmin	✔	✔	

How many children attend playgroup in the afternoon?

1 mark

Which child attends playgroup all day?

1 mark

3 A line measures 2 cm.

Draw a line that is three times longer.

Use a ruler.

1 mark

4 Dora has some place value cards.

300

400 500

30 7

60 10

2 5

Use the place value cards to create an **odd number between 378 and 416**

1 mark

5 Write the number **eight hundred and seven** in numerals.

1 mark

6 Ruth is facing the car.

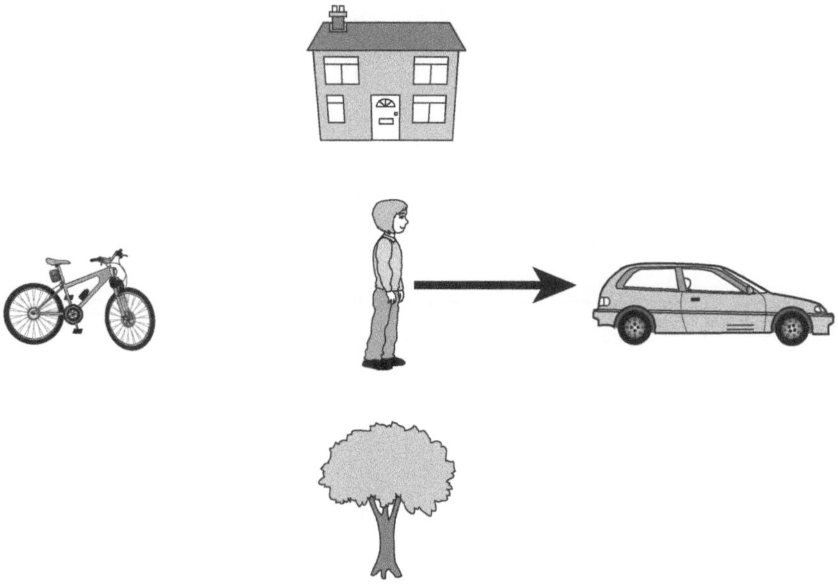

She turns to face the bicycle.

How many right angles does she turn through?

1 mark

7 Write the number **10 less** than 519.

[ ]

1 mark

8 Cross (**✗**) the incorrect calculation.

217 + 8 = 225	179 + 400 = 579
827 – 600 = 227	
717 – 40 = 687	509 – 20 = 489

1 mark

9 The table shows the number of people who visited an air show one weekend.

Saturday	398
Sunday	576

How many people visited the air show on this weekend altogether?

[ ]

1 mark

10 Sophia has made a model person using some 3-D shapes.

Match the parts of the body to the 3-D shape.

One is done for you.

Head	Sphere
Body	Cylinder
Arms	Cube
Legs	Cone

1 mark

11 Circle **two** calculations that have an answer of 48

4×8 6×8

4×12 3×12 8×12

6×12 3×4

1 mark

12 Estimate the number shown on the number line.

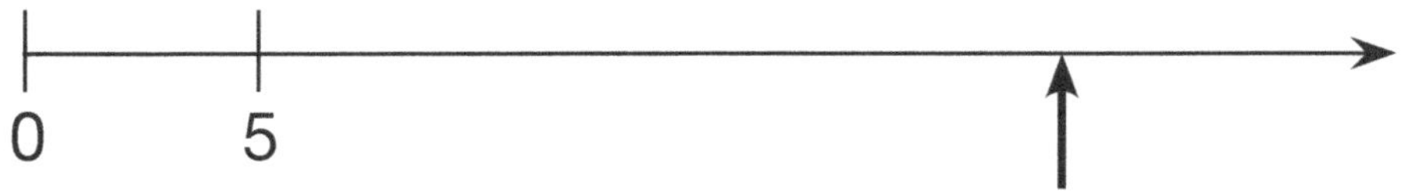

1 mark

13 Tick (✔) the shapes that have two pairs of parallel lines.

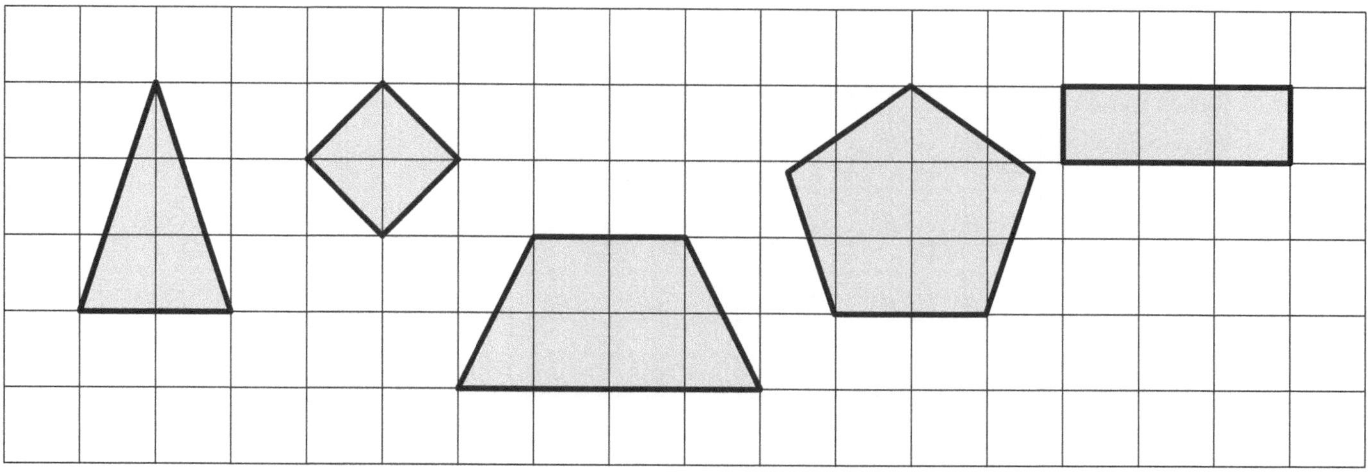

1 mark

14 Write a fraction in each box to complete the statement.

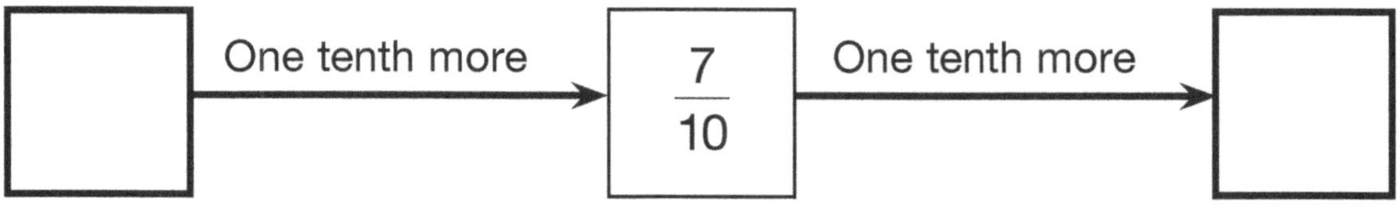

1 mark

15 The rectangle is drawn on a centimetre-squared grid.

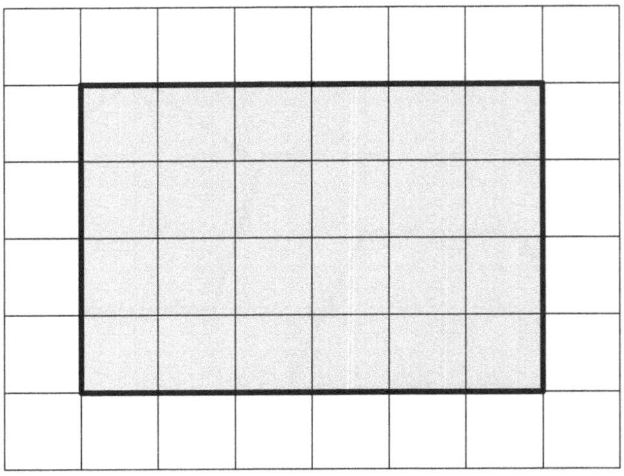

Calculate the perimeter of the rectangle.

cm

1 mark

16 Circle $\frac{2}{5}$ of these bees.

1 mark

17 The clock shows the time at 3 o'clock.

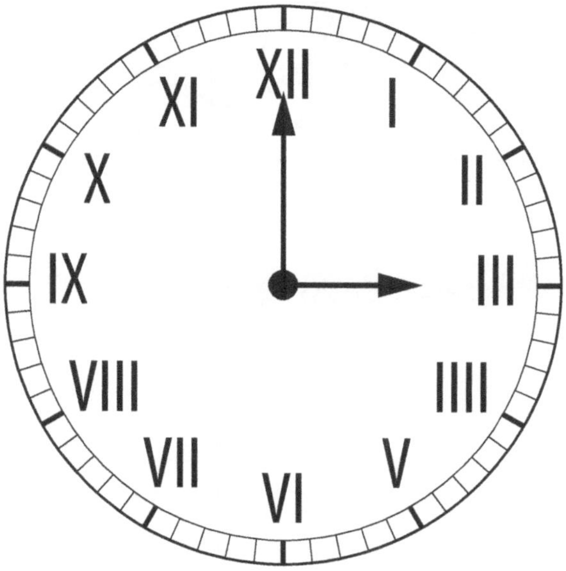

Draw the hands on the clock 45 minutes later.

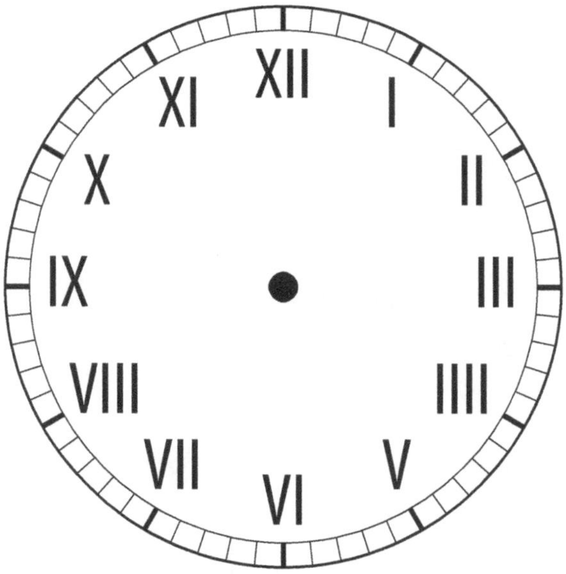

1 mark

18 Tennis balls are packed in cartons of 3

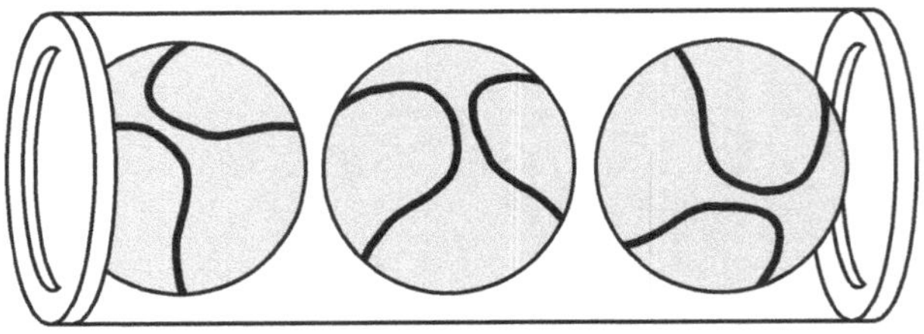

How many cartons are needed for 21 balls?

1 mark

19 The mass of a toy is 460 g

Draw the arrow on the scales to show the mass of the toy.

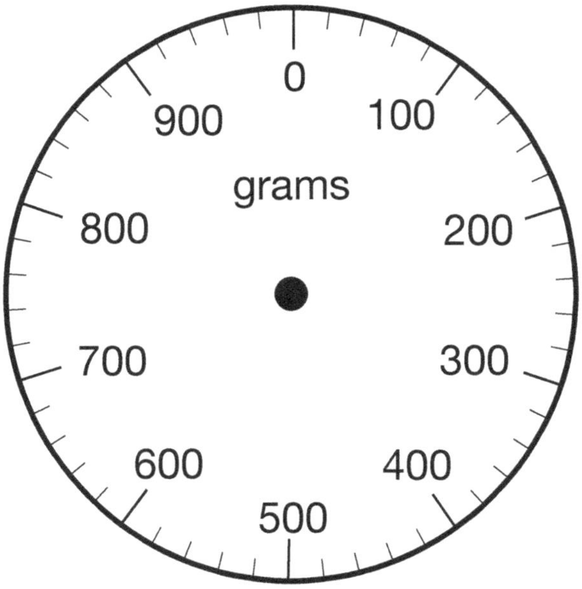

1 mark

20 Complete the calculation.

$$\frac{5}{8} + \frac{1}{8} + \square = 1$$

1 mark

21 The calendar for March 2015 is shown below.

Part of the calendar is missing.

March						
Mo	Tu	We	Th	Fr	Sa	Su
						1
2	3	4	5	6	7	8
9	10	11	12	13	14	15
16	17	18	19	20	21	22
23	24	25	26	27	28	29

What day of the week is 1st April 2015?

____

1 mark

22 How many numbers between 30 and 50 are multiples of 4?

1 mark

23 Jonti has 5 cups. He fills each cup with 250 ml of water.

Tick (✔) to show how much water is in the cups in total.

Less than 1 litre ☐

Exactly 1 litre ☐

More than 1 litre ☐

1 mark

24 The difference between the value of A and the value of B is 10.

What is the value of A?

[ ]

1 mark

25 A cake sale had 4 cakes for sale.

Each cake is cut into 8 slices.

Each slice is sold for £3.

What is the total amount earned for the sale of all the cakes?

Show your method

£ [ ]

2 marks

26 A gardener plants half her garden with potatoes.

She divided the other half into half to plant onions.

She divided the remaining section into three.

In one of the sections she grows tomatoes.

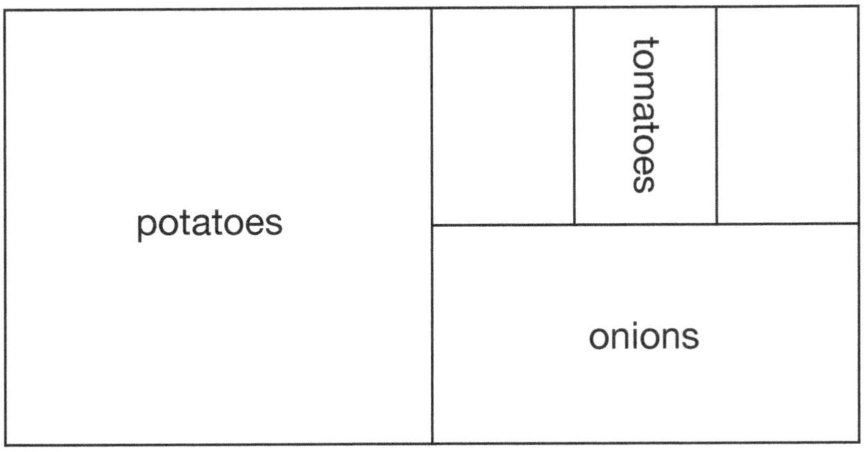

What fraction of the garden is used to grow tomatoes?

1 mark

27 **Use the digits 0, 1, 2, 3, 4 to complete the calculation.**

Use each digit only once.

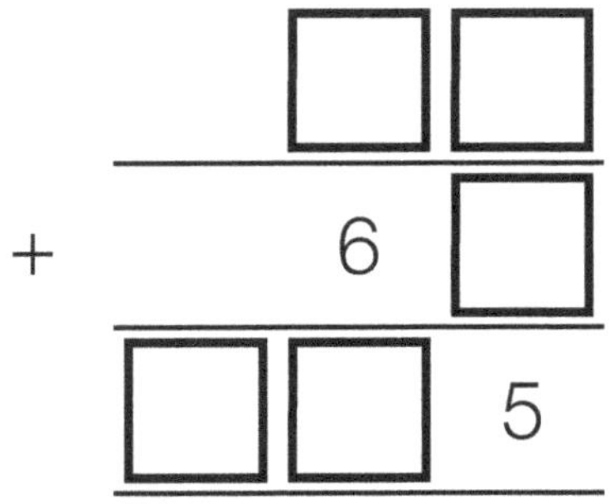

2 marks

Year 3
PAPER 3

Reasoning

Name: ______________________

Class: ______________________

Instructions

You **may not** use a calculator to answer any questions in this test.

Questions and answers

You have **30 minutes** to complete this test.

Follow the instructions for each question.

Work as quickly and as carefully as you can.

If you need to do working out, you can use the space around the question.

Some questions have a method box like this:

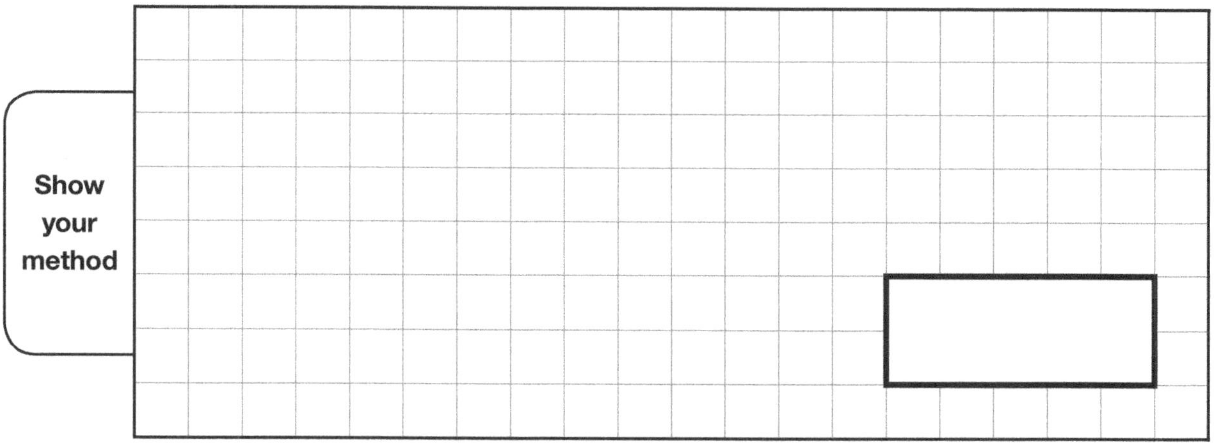

For these questions you may get a mark for showing your method.

If you cannot do one of the questions, **go to the next one**.
You can come back to it later, if you have time.

If you finish before the end, **go back and check your work**.

Marks

The number under each line at the side of the page tells you the maximum number of marks for each question.

1 Calculate the missing number.

$$400 + \square = 1000$$

1 mark

2 Draw a hexagon.

Use a ruler.

1 mark

3 James is playing a game with counters.

Each counter is worth a different amount.

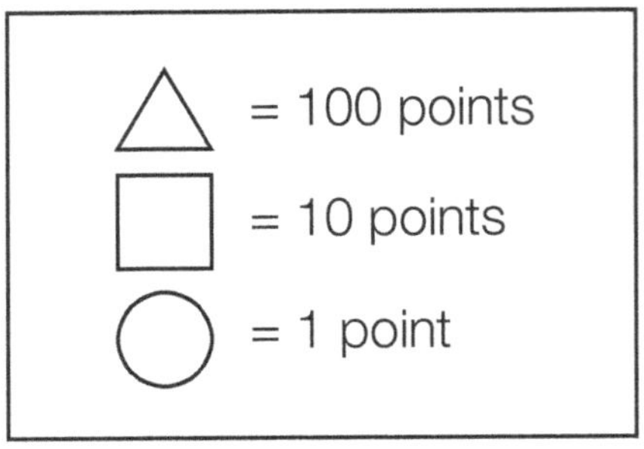

James has nine counters.

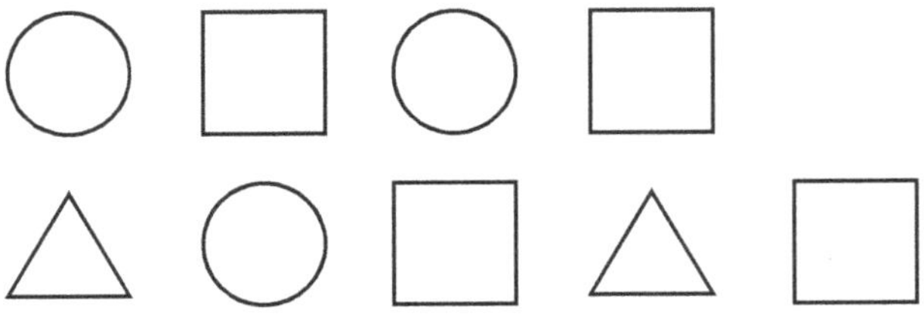

How many points does James have?

1 mark

Nesrin has 736 points.

She wins **one more triangular counter**.

How many points does she have now?

1 mark

4 Write the numbers in order from smallest to largest.

290 319 209 279 309

smallest largest

1 mark

5 Measure the longest side of the triangle.
Give your answer in millimetres.

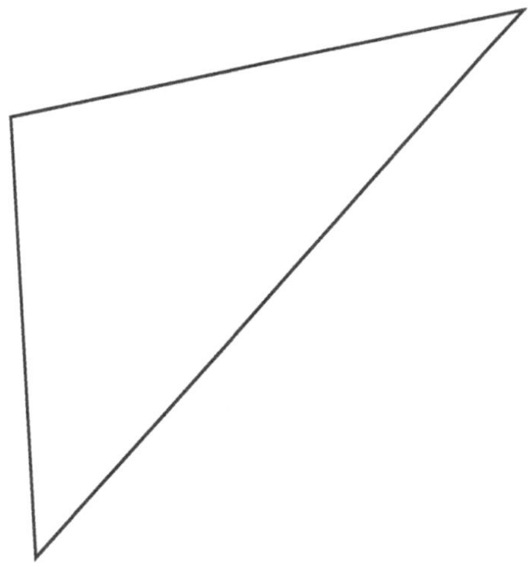

mm

1 mark

6 Mona is running an 800 m race.

She has run 545 m.

How many metres until the finish line?

Show your method

m

2 marks

7 The clock face shows the time in the evening.

Write this time using a.m. or p.m.

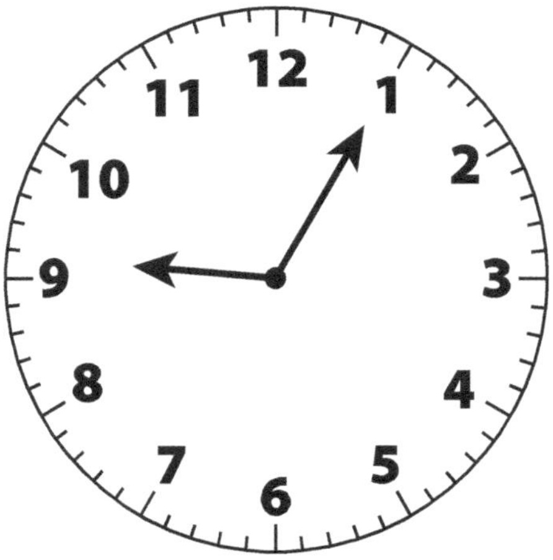

1 mark

8 Write the same number in each box to make the calculation correct.

$\square \times \square = 64$

1 mark

9 How much water is in the jug?

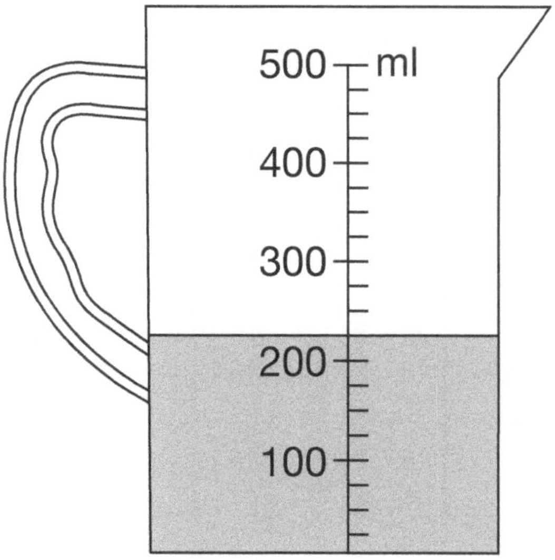

☐ ml

1 mark

10 The pictogram shows the times some children go to bed.

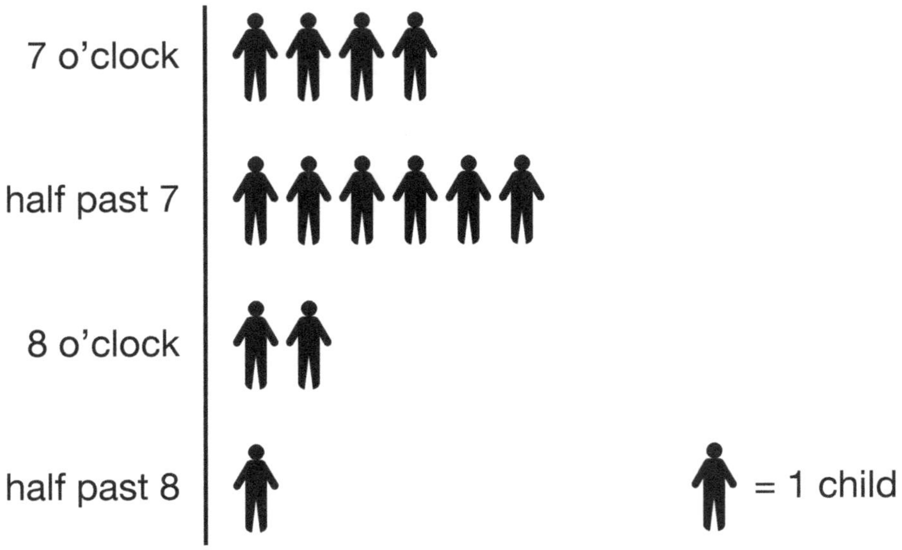

How many children go to bed at **8 o'clock or later**?

1 mark

11 Tick (✔) the shape that has no right angles.

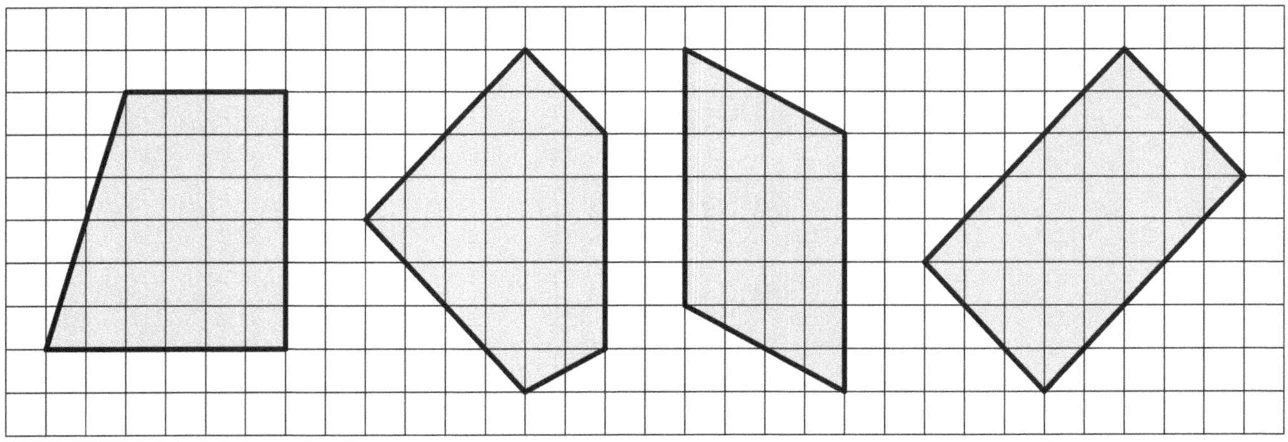

1 mark

12 Ellis used a balance to compare the masses of some objects.

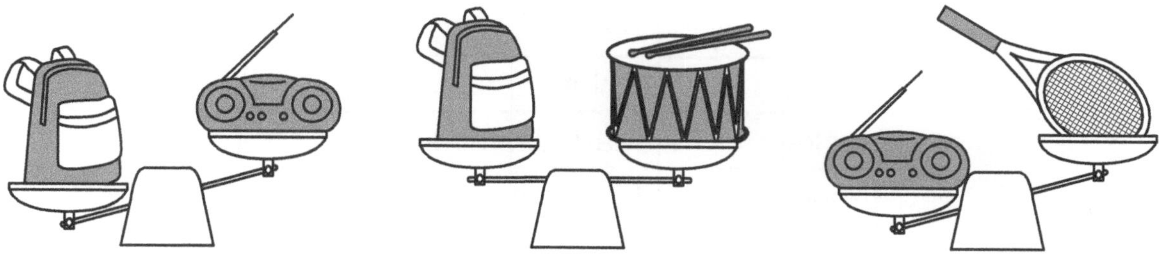

Tick (✔) the lightest object.

1 mark

13 Complete the calculation.

$8 \times 5 = 4 \times \square$

1 mark

14 Jacob collected some information about the sports played by his friends.

Cricket	Samir
Netball	Chloe, Eva, Layla
Hockey	Matt, Jasmine
Tennis	Phil, Zoe, Ali, Dina

Write the letter of the bar chart that shows Jacob's data.

A

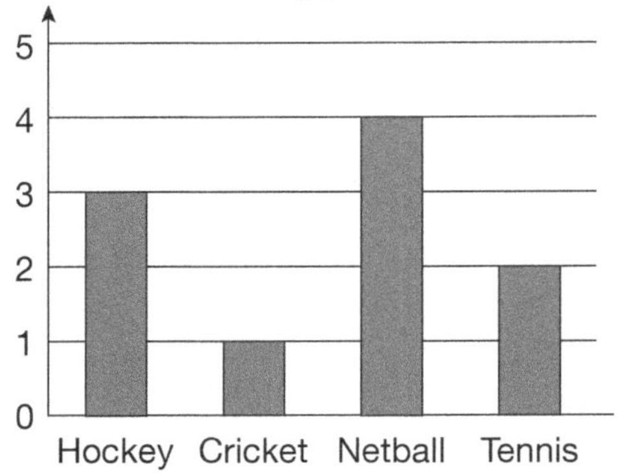

B

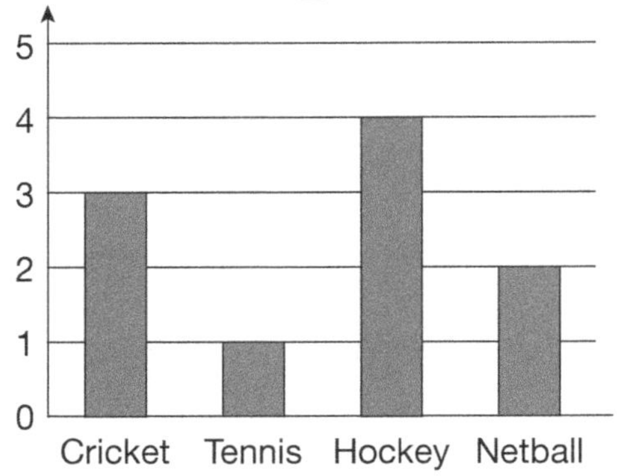

C

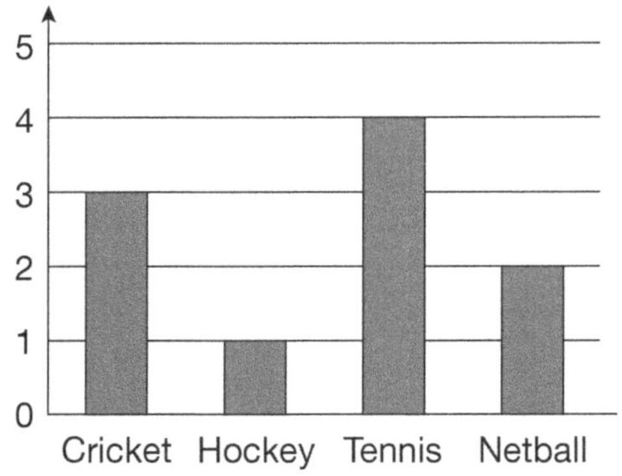

D

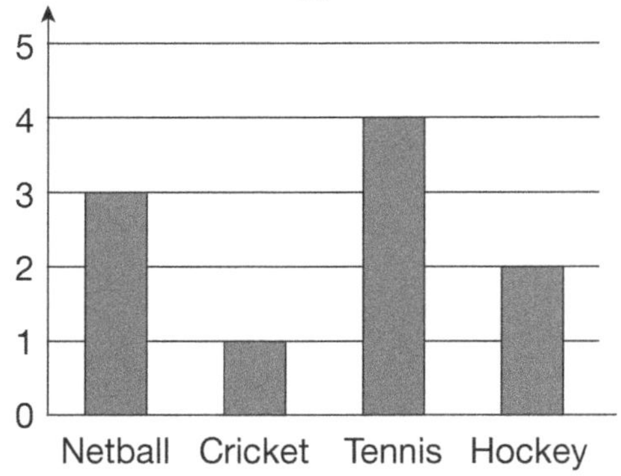

1 mark

15 Draw an arrow to show the position of $\frac{3}{10}$ on the number line.

1 mark

16 The times taken for 5 children to run a race are given in the table.

Name	Child
Ana	56 seconds
Barnaby	117 seconds
Caz	38 seconds
Dinesh	99 seconds
Eva	102 seconds

Tick (✔) the names of the children that ran the race in less than 1 minute.

Ana ☐ Barnaby ☐ Caz ☐

Dinesh ☐ Eva ☐

1 mark

17 The digit 9 can be made from 6 matchsticks.

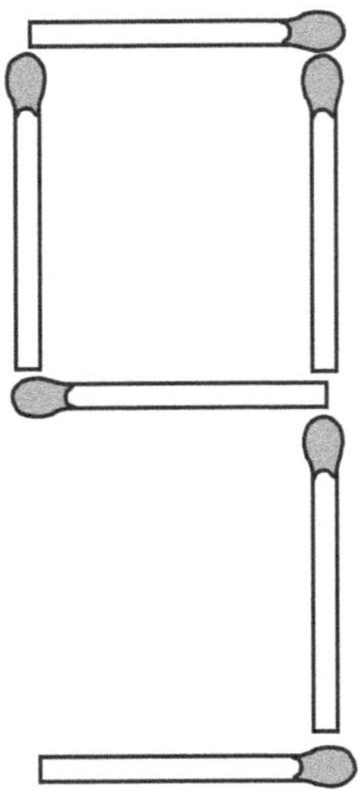

How many matchsticks are needed to make the number 99 999?

1 mark

How many digit 9s can be made from 66 matchsticks?

1 mark

18 One quarter of the regular octagon below is shaded.

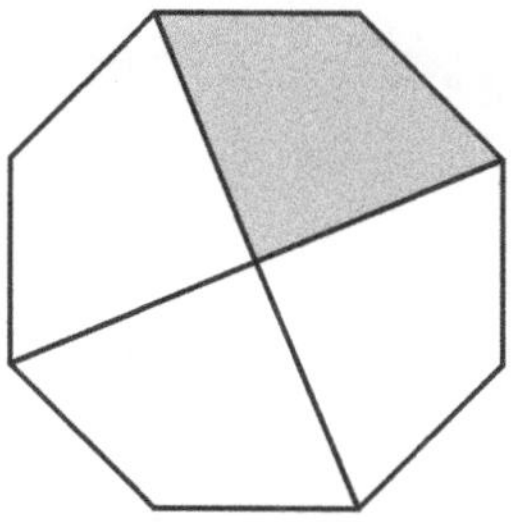

Tick (✔) the shape that has the same fraction shaded.

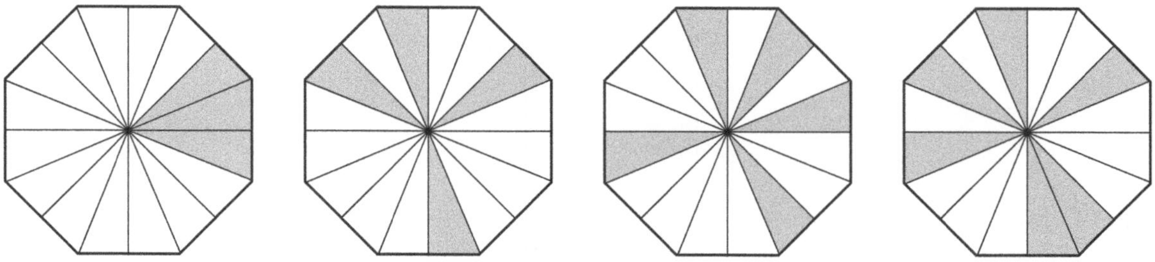

☐ ☐ ☐ ☐

1 mark

19 **Circle two numbers that add to 700**

325 385 345

440 355 360

1 mark

20 A pencil costs 38p.

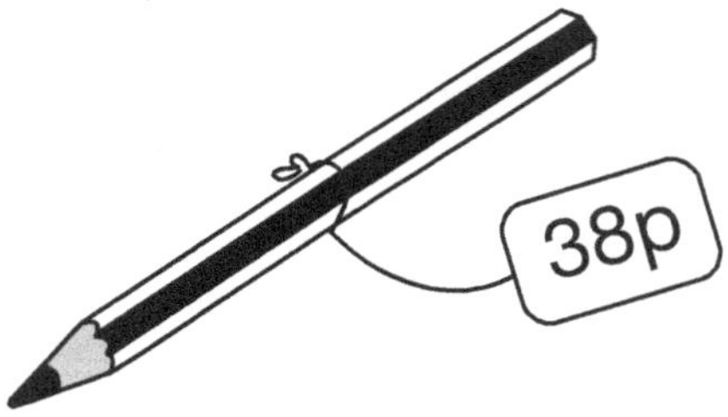

Sophie buys **two pencils**.

How much change does she get from £1?

Give your answer in pence.

Show your method.

Show your method

p

2 marks

21 Ella has a three-digit number.

She adds a one-digit number and the answer is 117

She subtracts the same one-digit number and the answer is 105

What is Ella's three-digit number?

1 mark

22 Complete each statement with < or >.

$\frac{1}{3}$ ☐ $\frac{1}{6}$ $\frac{4}{3}$ ☐ 1

$\frac{2}{9}$ ☐ $\frac{5}{9}$

1 mark

23 What is the first multiple of 8 greater than 500.

☐

1 mark

24 The number line below is divided into equal parts.

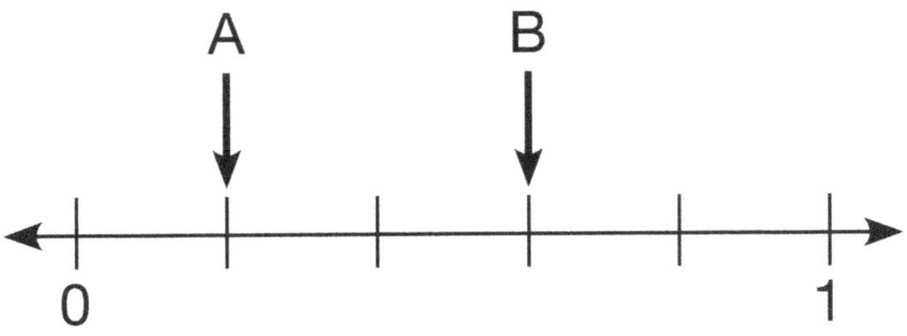

What is the difference between A and B on the number line?
Give your answer as a fraction.

1 mark

25 Sarah and Jake sell some boxes of cupcakes at the school fete.

There are 16 cakes in a box.

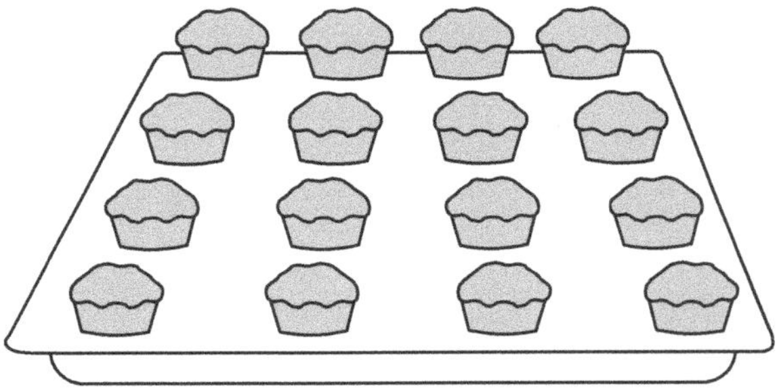

Sarah sells 5 boxes.

Jake sells 3 boxes.

Calculate how many **cakes** are sold altogether.

Show your method

2 marks

Year 3
PAPER
1

Arithmetic

Instructions

You **may not** use a calculator to answer any questions in this test.

Questions and answers

You have **25 minutes** to complete this test.

Work as quickly and as carefully as you can.

Put your answer in the box for each question.

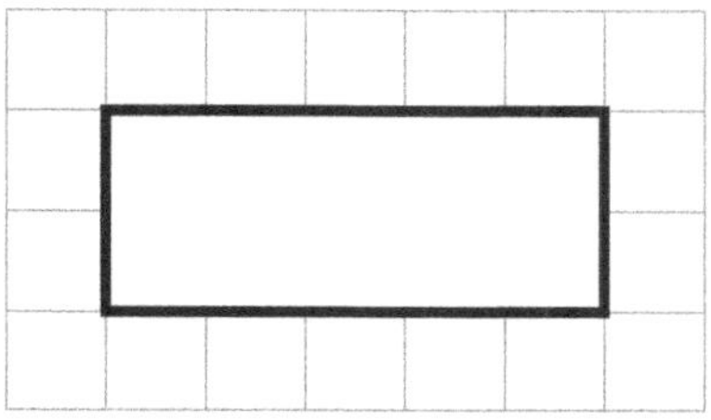

For questions expressed as common fractions, you should give your answer as common fractions.

All other answers should be given as whole numbers.

If you cannot do one of the questions, **go on to the next one**. You can come back to it later if you have time.

If you finish before the end, **go back and check your work**.

Marks

The number under each box at the side of the page tells you the maximum number of marks for each question.

In this test, some addition, subtraction and multiplication questions are worth **2 marks each**. You will be awarded 2 marks for a correct answer. You may get 1 mark for showing a formal method.

All other marks are worth **1 mark each**.

1 67 + 10 =

77

1 mark

2 44 − 7 =

37

1 mark

3 3 × 3 =

9

1 mark

4 658 – 40 =

618

1 mark

5 55 + 300 =

355

1 mark

6 576 – 100 =

476

1 mark

7 52 + 15 =

67

1 mark

8 178 + 9 =

187

1 mark

9 161 + 60 =

221

1 mark

10 $9 \times 4 =$

36

1 mark

11 $365 - 201 =$

164

1 mark

12 $\frac{2}{5} + \frac{1}{5} =$

$\frac{3}{5}$

1 mark

13 $16 \div 4 =$

4

1 mark

14 $\frac{1}{2}$ of 18 =

9

1 mark

15 $76 - 49 =$

27

1 mark

16 $550 + 50 =$

600

1 mark

17 $32 + 79 =$

111

1 mark

18 $33 \div 3 =$

11

1 mark

19 $12 \times 8 =$

96

1 mark

20 $854 + 62 =$

916

1 mark

21 $80 \div 4 =$

20

1 mark

22 $\frac{7}{9} - \frac{2}{9} =$

$\frac{5}{9}$

1 mark

23

$$\begin{array}{r} 2\,5\,6 \\ +\ 6\,6\,8 \\ \hline \end{array}$$

Show your method

924

2 marks

24

$$\begin{array}{r} 2\,4 \\ \times\ \ 9 \\ \hline \end{array}$$

Show your method

216

2 marks

25

$$\begin{array}{r} 753 \\ -\ 177 \\ \hline \end{array}$$

Show your method

576

2 marks

26

$96 \div 3 =$

32

1 mark

27

$\frac{4}{5}$ of 25 =

20

1 mark

Year 3
PAPER
2

Reasoning

Instructions

You **may not** use a calculator to answer any questions in this test.

Questions and answers

You have **30 minutes** to complete this test.

Follow the instructions for each question.

Work as quickly and as carefully as you can.

If you need to do working out, you can use the space around the question.

Some questions have a method box like this:

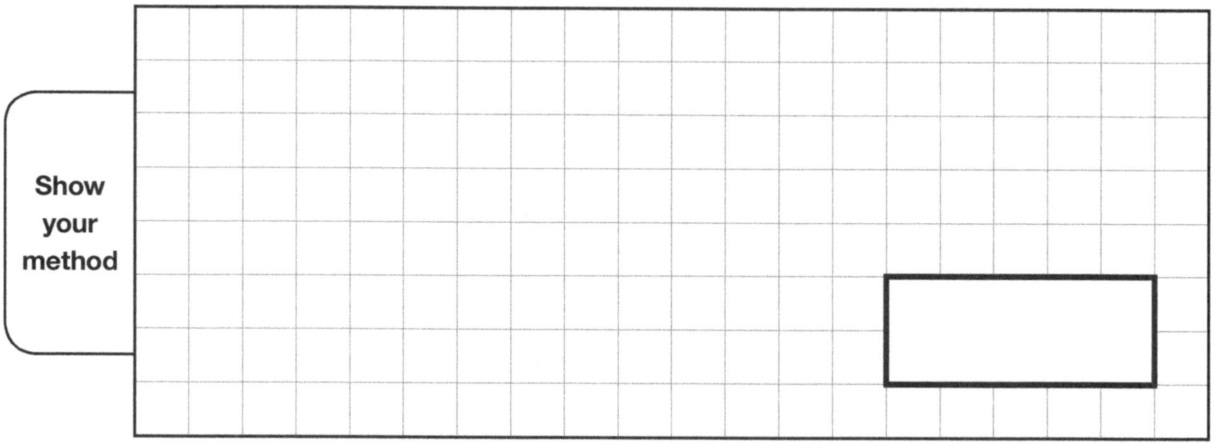

For these questions you may get a mark for showing your method.

If you cannot do one of the questions, **go to the next one**.
You can come back to it later, if you have time.

If you finish before the end, **go back and check your work**.

Marks

The number under each line at the side of the page tells you the maximum number of marks for each question.

1 Write the missing number.

400	500	600	700

1 mark

2 Children can attend playgroup for the morning, lunch and afternoon.

The sessions attended by 6 children are shown in the table.

	Morning	Lunch	Afternoon
Ava	✔	✔	
Grace	✔	✔	✔
Jayden		✔	✔
Kai	✔		
Frankie			✔
Yasmin	✔	✔	

How many children attend playgroup in the afternoon?

3

1 mark

Which child attends playgroup all day?

Grace

1 mark

3 A line measures 2 cm.

Draw a line that is three times longer.

Use a ruler.

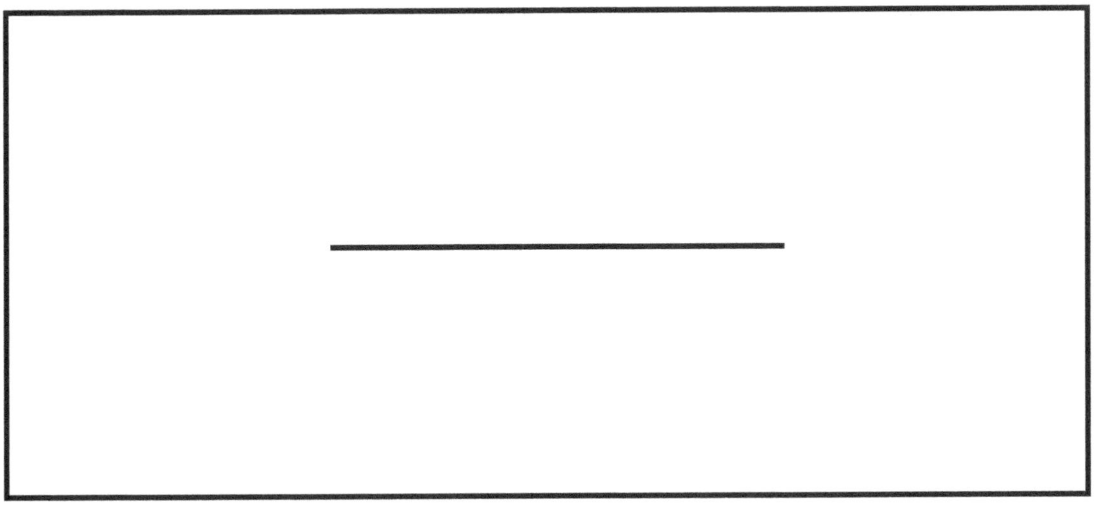

1 mark

4 Dora has some place value cards.

300

400 500

30 7

60 10

2 5

Use the place value cards to create an **odd number between 378 and 416**

4	1	5

1 mark

5 Write the number **eight hundred and seven** in numerals.

807

1 mark

6 Ruth is facing the car.

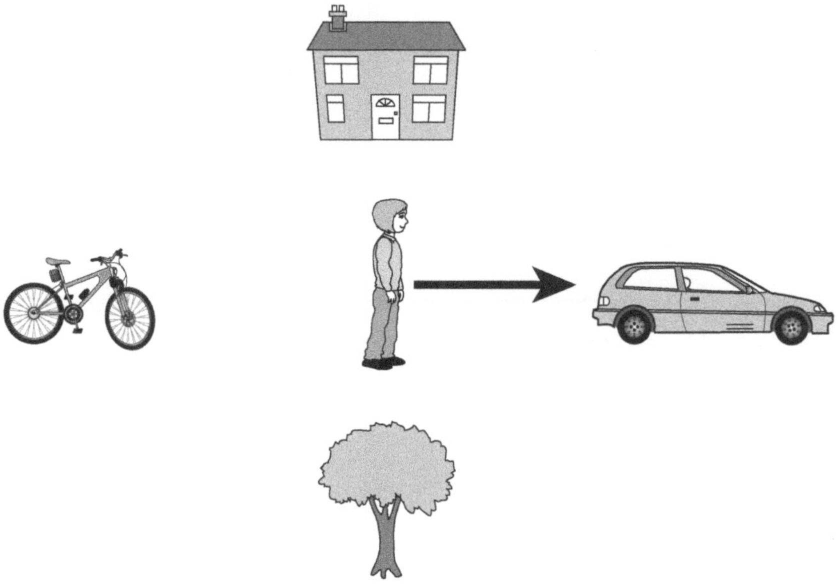

She turns to face the bicycle.

How many right angles does she turn through?

2

1 mark

7 Write the number **10 less** than 519.

509

1 mark

8 Cross (✗) the incorrect calculation.

217 + 8 = 225	179 + 400 = 579
827 – 600 = 227	
717 – ✗ = 687	509 – 20 = 489

1 mark

9 The table shows the number of people who visited an air show one weekend.

Saturday	398
Sunday	576

How many people visited the air show on this weekend altogether?

974

1 mark

10 Sophia has made a model person using some 3-D shapes.

Match the parts of the body to the 3-D shape.

One is done for you.

Part of body	3-D shape
Head	Sphere
Body	Cylinder
Arms	Cube
Legs	Cone

1 mark

11 Circle **two** calculations that have an answer of 48

4 × 8   6 × 8

4 × 12   3 × 12   8 × 12

6 × 12   3 × 4

1 mark

12 Estimate the number shown on the number line.

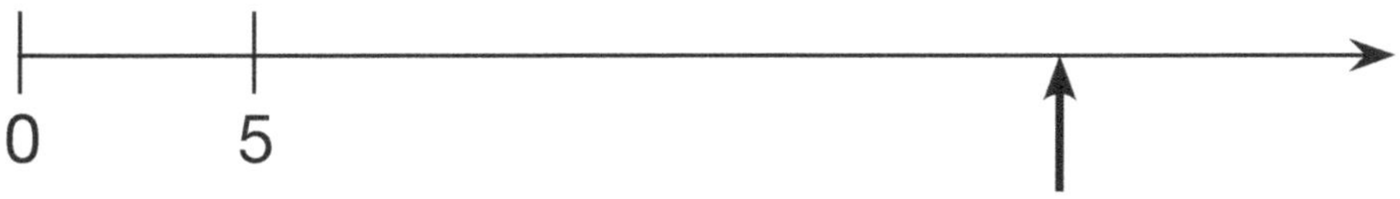

20-24

1 mark

13 Tick (✔) the shapes that have two pairs of parallel lines.

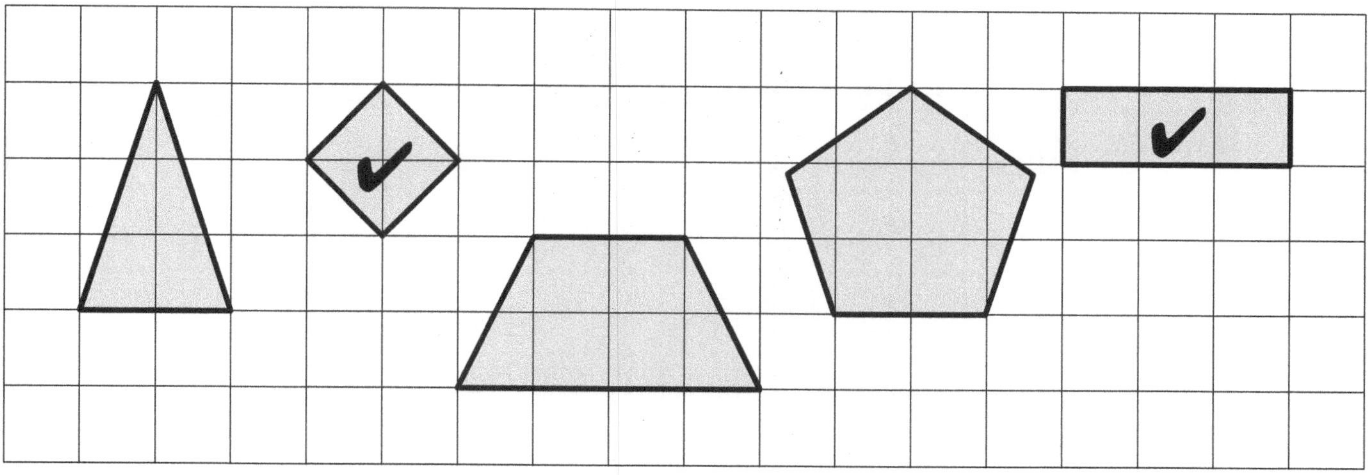

1 mark

14 Write a fraction in each box to complete the statement.

$\frac{6}{10}$ → One tenth more → $\frac{7}{10}$ → One tenth more → $\frac{8}{10}$

1 mark

15 The rectangle is drawn on a centimetre-squared grid.

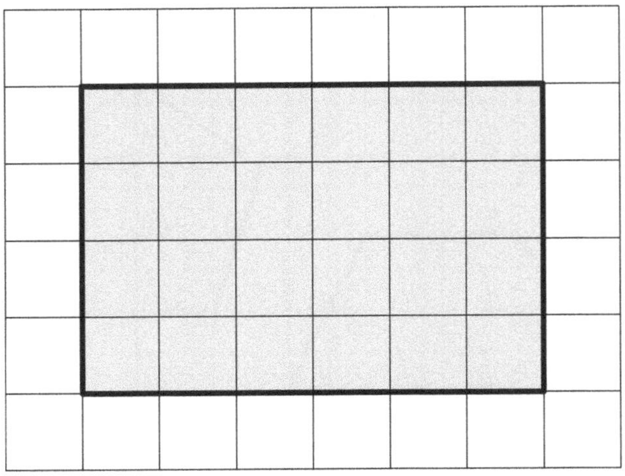

Calculate the perimeter of the rectangle.

20 cm

1 mark

16 Circle $\frac{2}{5}$ of these bees.

1 mark

17 The clock shows the time at 3 o'clock.

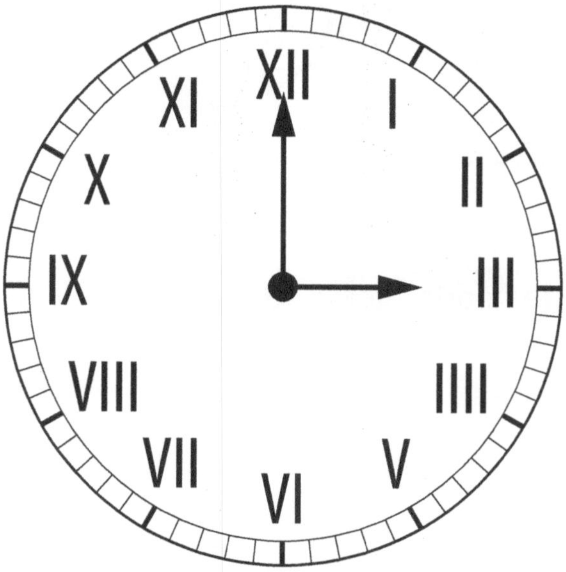

Draw the hands on the clock 45 minutes later.

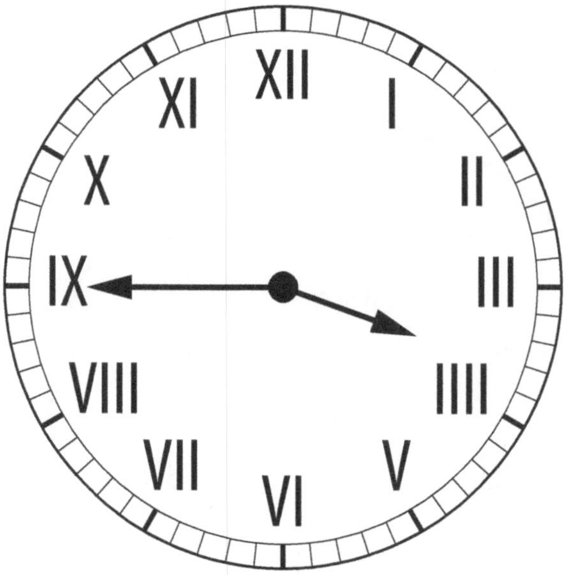

1 mark

18 Tennis balls are packed in cartons of 3

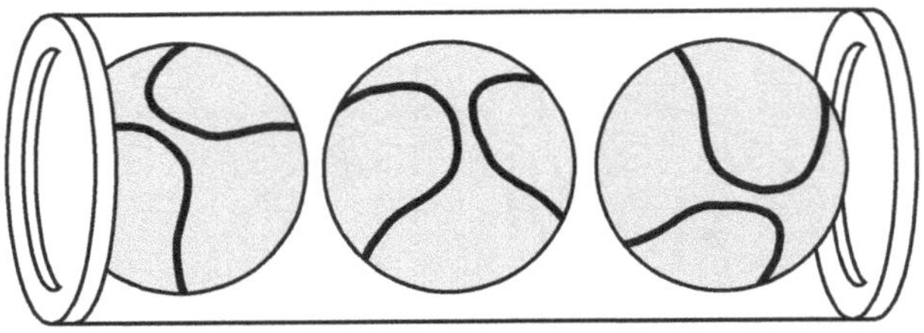

How many cartons are needed for 21 balls?

7

1 mark

19 The mass of a toy is 460 g

Draw the arrow on the scales to show the mass of the toy.

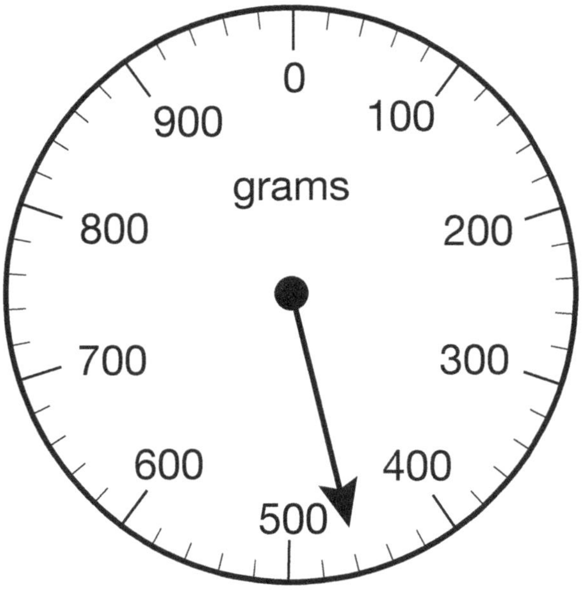

1 mark

20 Complete the calculation.

$$\frac{5}{8} + \frac{1}{8} + \boxed{\frac{2}{8}} = 1$$

1 mark

21 The calendar for March 2015 is shown below.

Part of the calendar is missing.

March						
Mo	Tu	We	Th	Fr	Sa	Su
						1
2	3	4	5	6	7	8
9	10	11	12	13	14	15
16	17	18	19	20	21	22
23	24	25	26	27	28	29

What day of the week is 1st April 2015?

Wednesday

1 mark

22 How many numbers between 30 and 50 are multiples of 4?

5

1 mark

23 Jonti has 5 cups. He fills each cup with 250 ml of water.

Tick (✔) to show how much water is in the cups in total.

Less than 1 litre ☐

Exactly 1 litre ☐

More than 1 litre ☑

1 mark

24 The difference between the value of A and the value of B is 10.

What is the value of A?

4

1 mark

25 A cake sale had 4 cakes for sale.

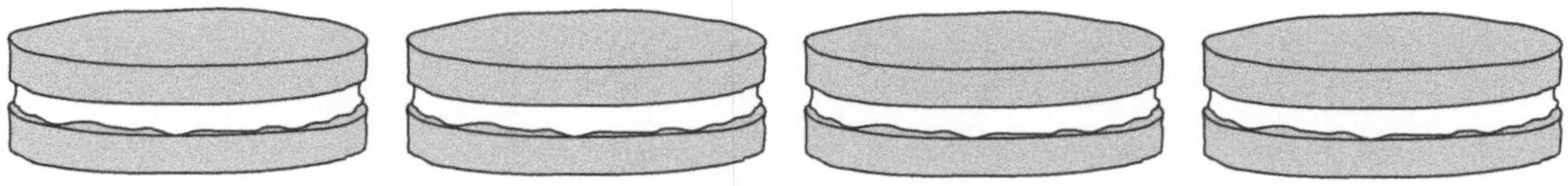

Each cake is cut into 8 slices.

Each slice is sold for £3.

What is the total amount earned for the sale of all the cakes?

Show your method

£ 96

2 marks

26 A gardener plants half her garden with potatoes.

She divided the other half into half to plant onions.

She divided the remaining section into three.

In one of the sections she grows tomatoes.

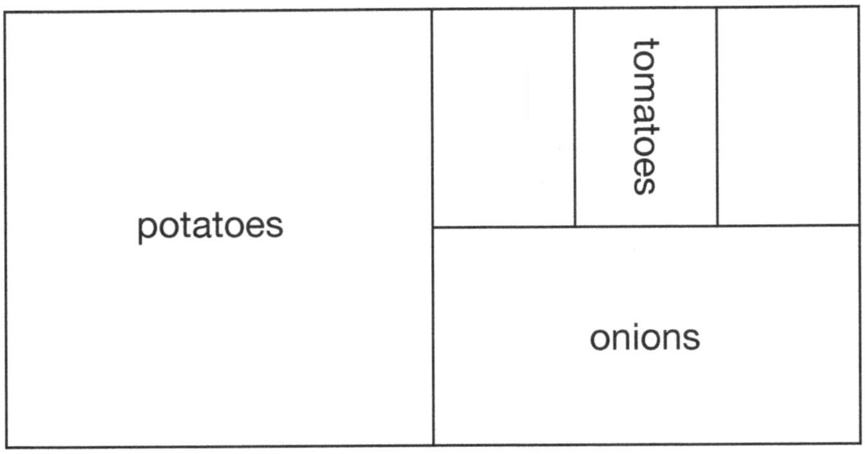

What fraction of the garden is used to grow tomatoes?

$\frac{1}{12}$

1 mark

27 **Use the digits 0, 1, 2, 3, 4 to complete the calculation.**

Use each digit only once.

		4	2				4	3
+		6	3	or	+		6	2
	1	0	5			1	0	5

2 marks

Year 3
PAPER
3

Reasoning

Instructions

You **may not** use a calculator to answer any questions in this test.

Questions and answers

You have **30 minutes** to complete this test.

Follow the instructions for each question.

Work as quickly and as carefully as you can.

If you need to do working out, you can use the space around the question.

Some questions have a method box like this:

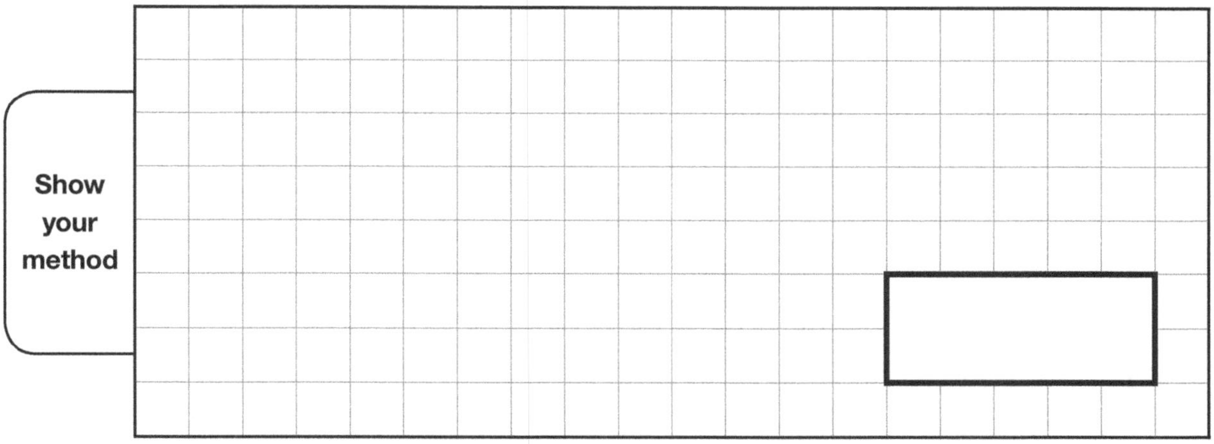

For these questions you may get a mark for showing your method.

If you cannot do one of the questions, **go to the next one**.
You can come back to it later, if you have time.

If you finish before the end, **go back and check your work**.

Marks

The number under each line at the side of the page tells you the maximum number of marks for each question.

1 Calculate the missing number.

$$400 + \boxed{600} = 1000$$

1 mark

2 Draw a hexagon.

Use a ruler.

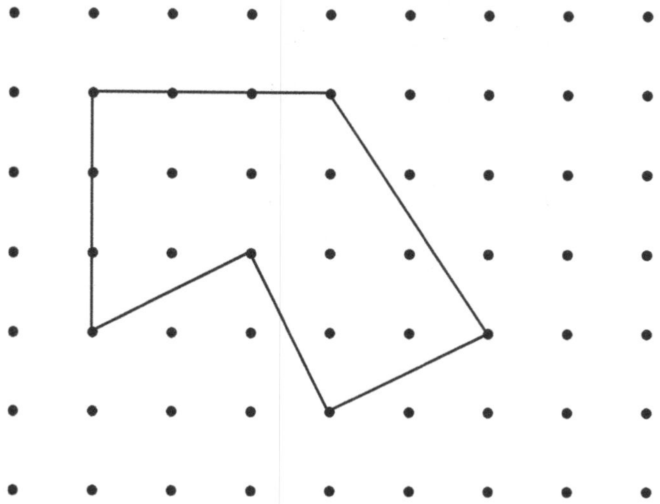

1 mark

3 James is playing a game with counters.

Each counter is worth a different amount.

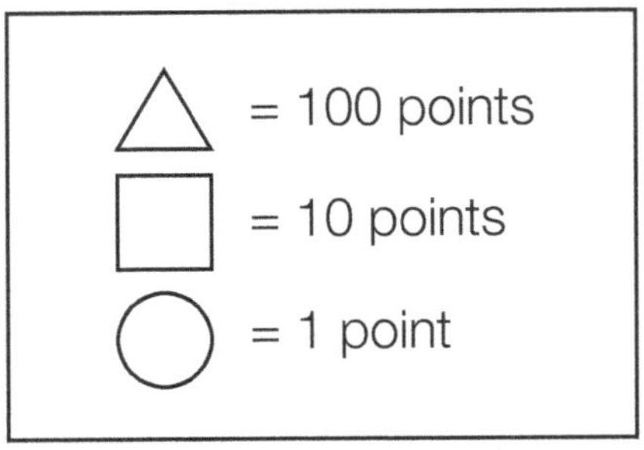

James has nine counters.

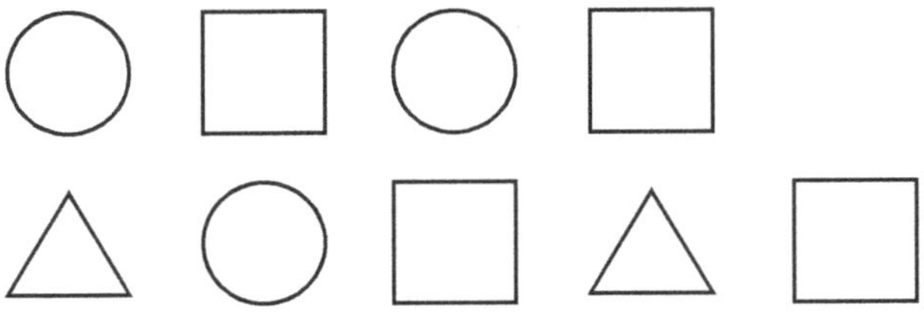

How many points does James have?

243

1 mark

Nesrin has 736 points.

She wins **one more triangular counter**.

How many points does she have now?

836

1 mark

4 Write the numbers in order from smallest to largest.

290 319 209 279 309

209	279	290	309	319

smallest ... largest

1 mark

5 Measure the longest side of the triangle.
Give your answer in millimetres.

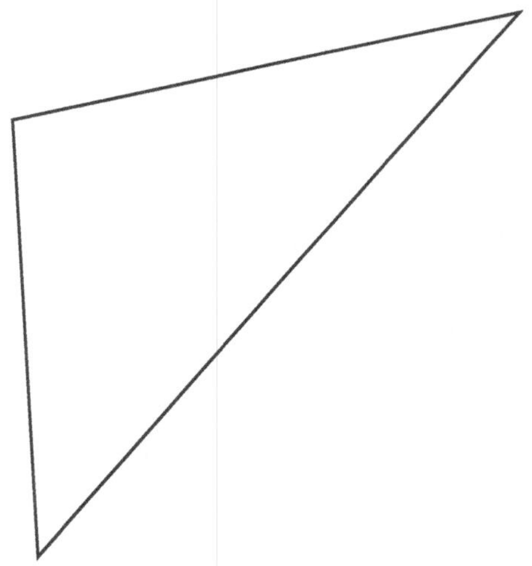

81-85 mm

1 mark

6 Mona is running an 800 m race.

She has run 545 m.

How many metres until the finish line?

Show your method

255 m

2 marks

7 The clock face shows the time in the evening.

Write this time using a.m. or p.m.

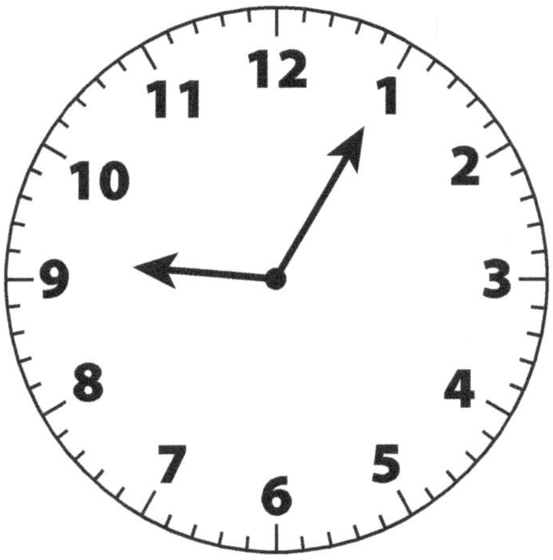

9:05 p.m.

1 mark

8 Write the same number in each box to make the calculation correct.

8 × 8 = 64

1 mark

9 How much water is in the jug?

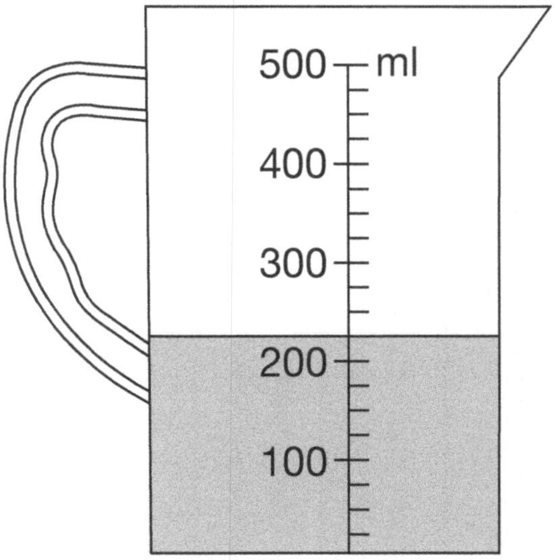

225 ml

1 mark

10 The pictogram shows the times some children go to bed.

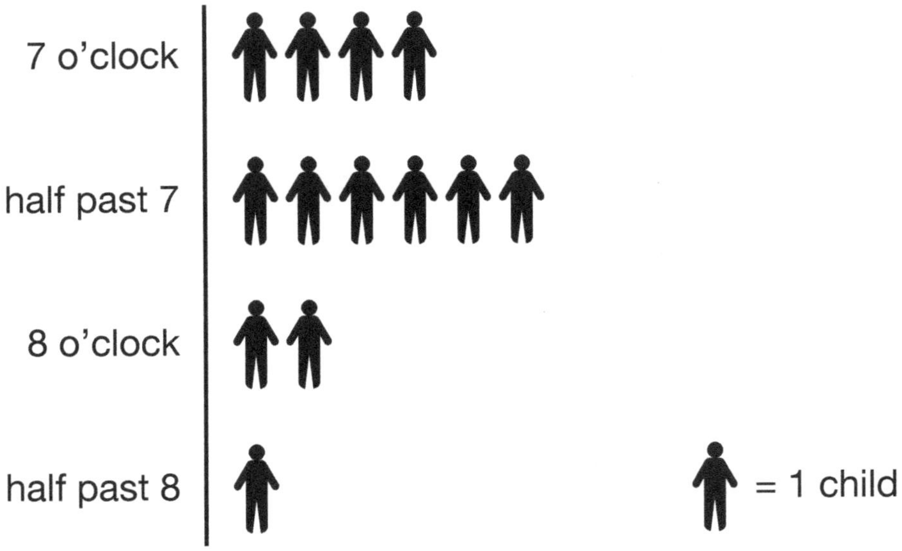

How many children go to bed at **8 o'clock or later**?

3

1 mark

11 Tick (✔) the shape that has no right angles.

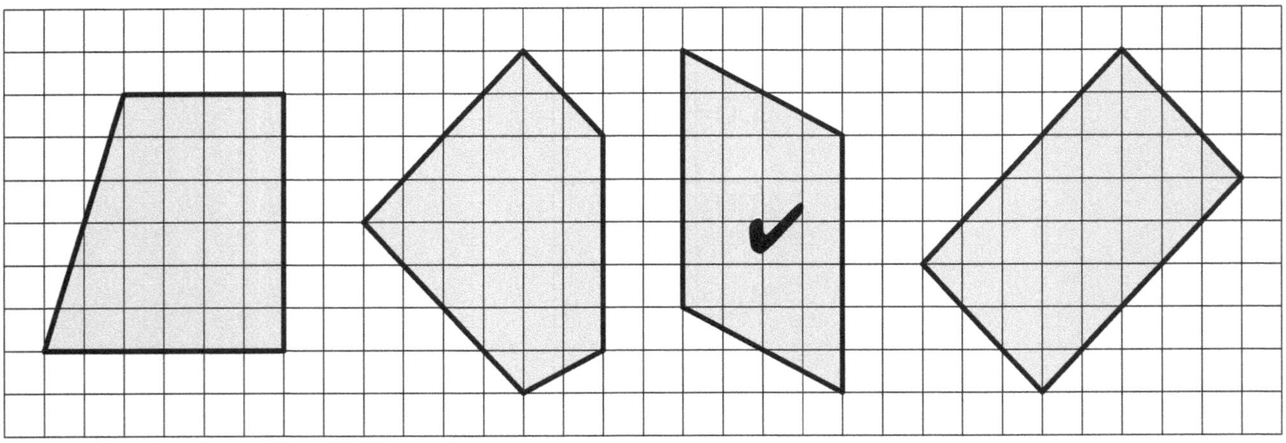

1 mark

12 Ellis used a balance to compare the masses of some objects.

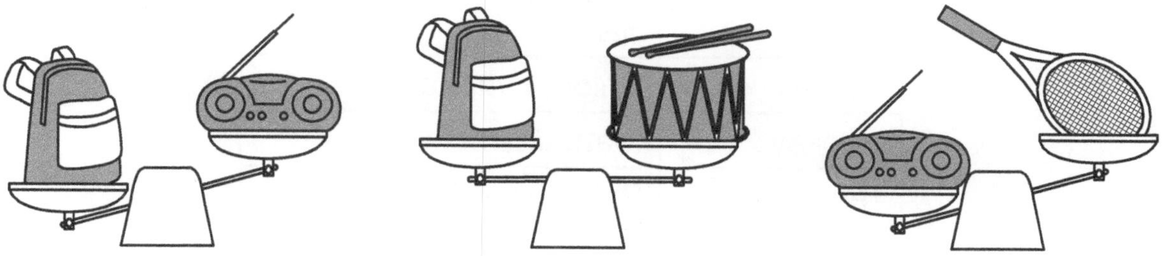

Tick (✔) the lightest object.

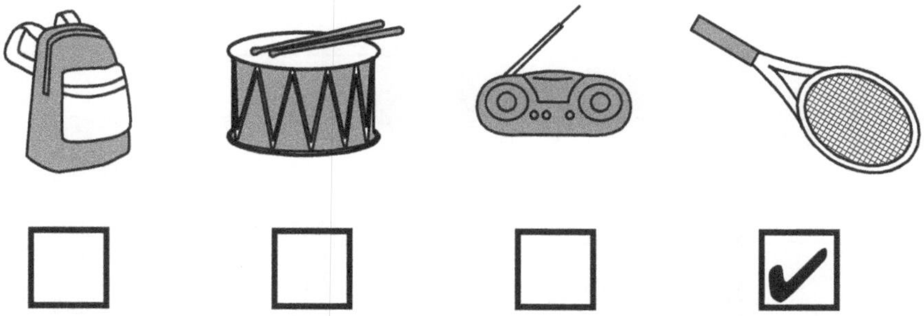

1 mark

13 Complete the calculation.

$8 \times 5 = 4 \times$ 10

1 mark

14 Jacob collected some information about the sports played by his friends.

Cricket	Samir
Netball	Chloe, Eva, Layla
Hockey	Matt, Jasmine
Tennis	Phil, Zoe, Ali, Dina

Write the letter of the bar chart that shows Jacob's data.

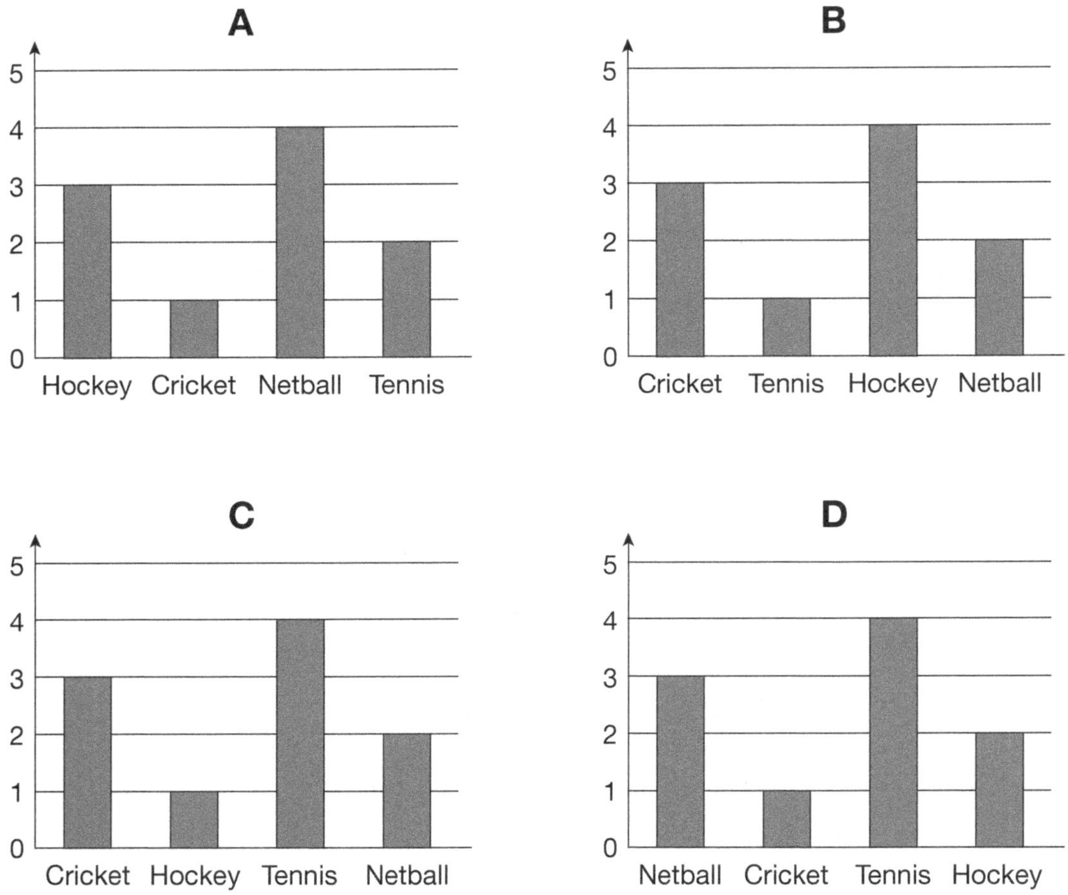

D

1 mark

15 Draw an arrow to show the position of $\frac{3}{10}$ on the number line.

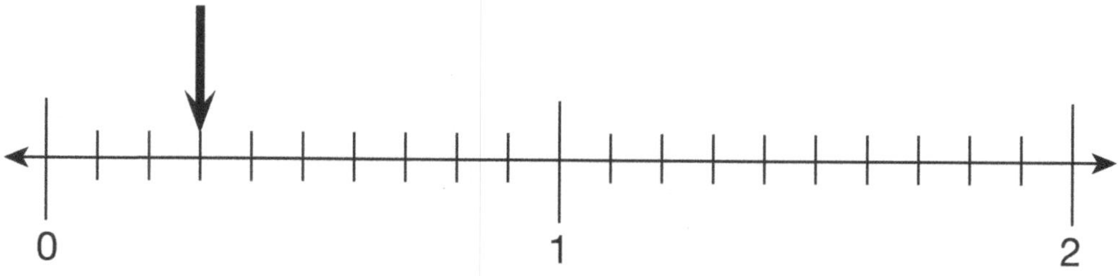

1 mark

16 The times taken for 5 children to run a race are given in the table.

Name	Child
Ana	56 seconds
Barnaby	117 seconds
Caz	38 seconds
Dinesh	99 seconds
Eva	102 seconds

Tick (✔) the names of the children that ran the race in less than 1 minute.

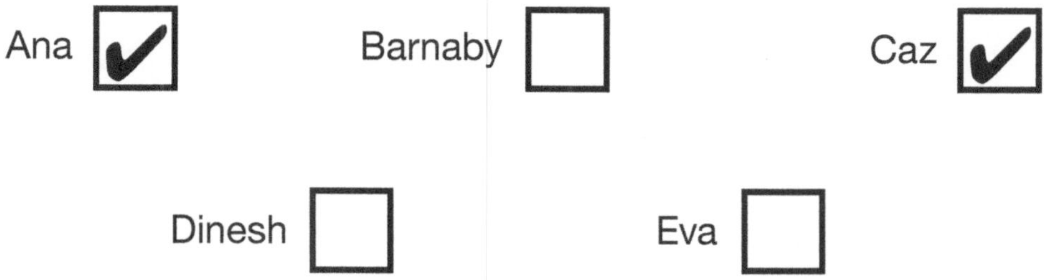

1 mark

17 The digit 9 can be made from 6 matchsticks.

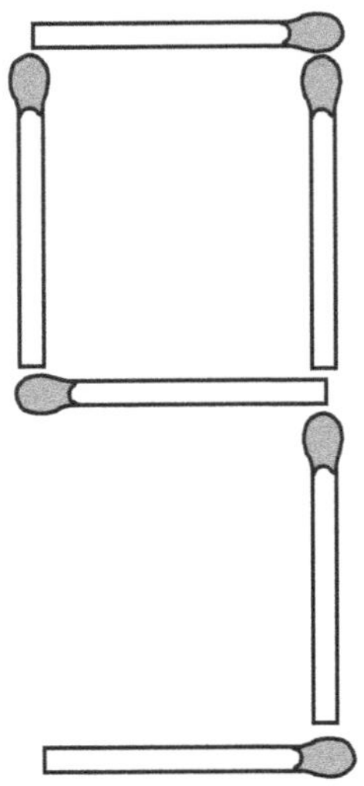

How many matchsticks are needed to make the number 99 999?

30

1 mark

How many digit 9s can be made from 66 matchsticks?

11

1 mark

18 One quarter of the regular octagon below is shaded.

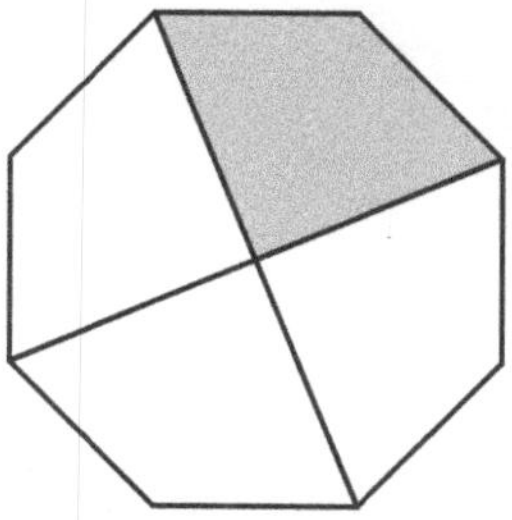

Tick (✔) the shape that has the same fraction shaded.

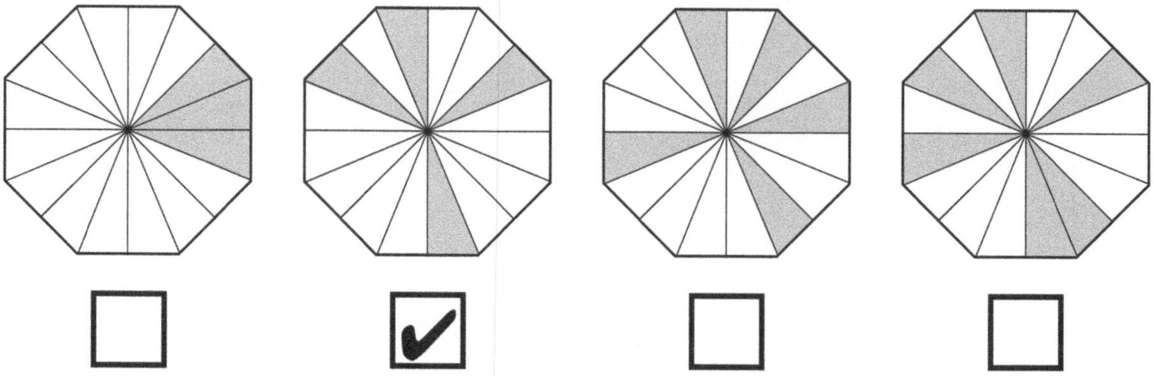

1 mark

19 Circle **two** numbers that add to 700

325 385 (345)

440 (355) 360

1 mark

20 A pencil costs 38p.

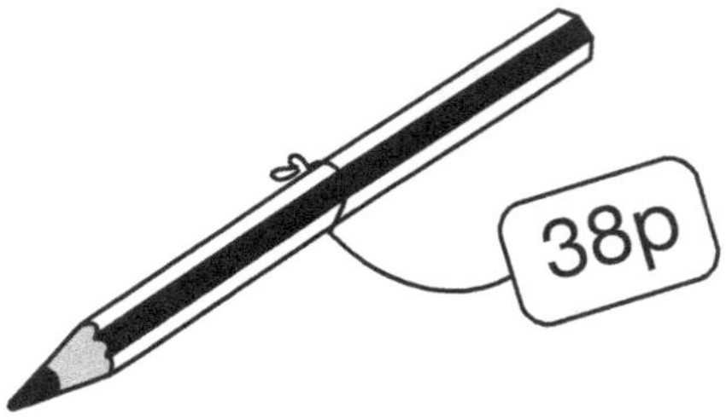

Sophie buys **two pencils**.

How much change does she get from £1?

Give your answer in pence.

Show your method.

Show your method

24 p

2 marks

21 Ella has a three-digit number.

She adds a one-digit number and the answer is 117

She subtracts the same one-digit number and the answer is 105

What is Ella's three-digit number?

1	1	1

1 mark

22 Complete each statement with < or >.

$\frac{1}{3}$ [>] $\frac{1}{6}$

$\frac{4}{3}$ [>] 1

$\frac{2}{9}$ [<] $\frac{5}{9}$

1 mark

23 What is the first multiple of 8 greater than 500.

[504]

1 mark

24 The number line below is divided into equal parts.

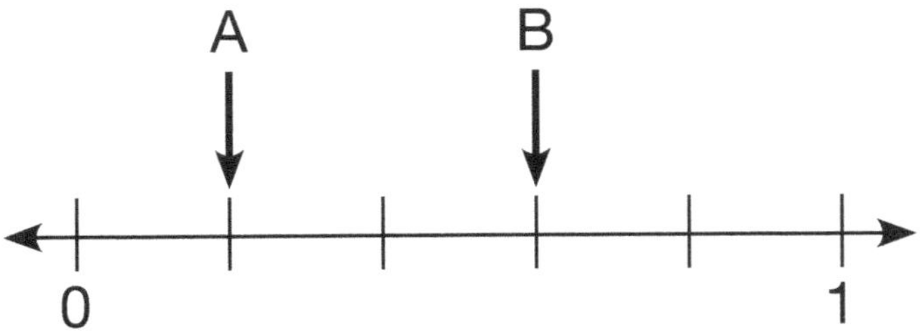

What is the difference between A and B on the number line?
Give your answer as a fraction.

$\frac{2}{5}$

1 mark

25 Sarah and Jake sell some boxes of cupcakes at the school fete.

There are 16 cakes in a box.

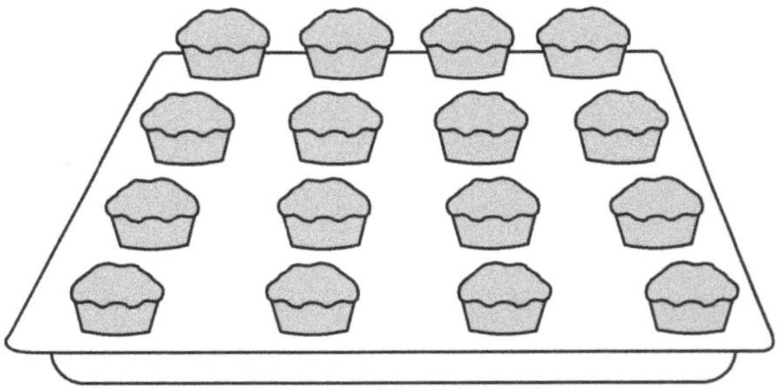

Sarah sells 5 boxes.

Jake sells 3 boxes.

Calculate how many **cakes** are sold altogether.

Show your method

128

2 marks